Peter Hölz

A dynamic and statistical analysis of the temperature- and fatigue behavior of a race power unit

The effect of different thermodynamic states

Schriftenreihe
Kontinuumsmechanik im Maschinenbau
Band 16

Karlsruher Institut für Technologie (KIT)
Institut für Technische Mechanik
Bereich Kontinuumsmechanik

Hrsg. Prof. Dr.-Ing. habil. Thomas Böhlke

Eine Übersicht aller bisher in dieser Schriftenreihe erschienenen Bände finden Sie am Ende des Buchs.

A dynamic and statistical analysis of the temperature- and fatigue behavior of a race power unit

The effect of different thermodynamic states

by
Peter Hölz

KIT Scientific Publishing

Karlsruher Institut für Technologie
Institut für Technische Mechanik, Bereich Kontinuumsmechanik

A dynamic and statistical analysis of the temperature- and fatigue behavior of a race power unit – The effect of different thermodynamic states

Zur Erlangung des akademischen Grades eines Doktor-Ingenieurs von der KIT-Fakultät für Maschinenbau des Karlsruher Instituts für Technologie (KIT) genehmigte Dissertation

von M.Sc. Peter Hölz

Tag der mündlichen Prüfung: 11. September 2019
Referenten: Prof. Dr.-Ing. Thomas Böhlke
Prof. Dr.-Ing. Federico Millo
Prof. Dr.-Ing. Thomas Seifert

Impressum

KIT Scientific Publishing

Karlsruher Institut für Technologie (KIT)
KIT Scientific Publishing
Straße am Forum 2
D-76131 Karlsruhe

KIT Scientific Publishing is a registered trademark of Karlsruhe Institute of Technology.
Reprint using the book cover is not allowed.

www.ksp.kit.edu

Print on Demand 2020 – Gedruckt auf FSC-zertifiziertem Papier

ISSN 2192-693X
ISBN 978-3-7315-0988-2
DOI 10.5445/KSP/1000099193

Posthumous recognition of the scientific career of Dr. Ing. Peter Hölz

† December 27, 2019

We would like to express our most sincere sympathies to the family of Peter Hölz and recognize, herewith, his extraordinary talent and extremely bright knowledge shown during his studies and research activities at the Karlsruhe Institute of Technologies (KIT) as well as doctoral researcher supported by the Dr. Ing. h.c. F. Porsche AG.

In 2008, the year he very successfully finished the high school in Riedlingen, Germany, Peter Hölz started with his bachelor studies in Mechanical Engineering at KIT. From January 2010 until May 2011, he had been a member of the Formula Student Team „KA-RaceIng", (construction/simulation of landing gear). From October 2009 until February 2010, he had been employed as tutor at the Institute for Information Management in Engineering at KIT, and from January 2011 until July 2011, Peter Hölz had been assistant scientist in the field of mechanics of fiber composites at the Institute of Engineering Mechanics at KIT. He finished his bachelor studies in 2012 with the excellent bachelor thesis „Fracture modeling of fiber reinforced composite materials – analysis and FEM implementation of the Puck model" at the Institute of Engineering Mechanics at KIT (Prof. Böhlke) in cooperation with Suisse Technology Partners, Neuhausen, Switzerland. Peter Hölz was an outstanding, dedicated student not only with a keen perception, but also with the extreme capability to scrutinize what he had learned. During his courses on thermodynamics and combustion processes he pointed out several inaccuracies and errors in the textbook „Combustion", and thus helped considerably to improve this textbook. Furthermore, he followed all the courses he visited with intensity, not only interest but pure joy, with valuable feedback also for the lecturer and positive influence on the complete group.

Based on his outstanding grades in his bachelor studies (final grade 1.2), he was awarded a Scholarship for gifted students by the Gunther-Schroff-Stiftung in Germany in 2010 which was followed by the award of receiving the Grashof prize of the faculty of Mechanical Engineering of KIT in 2011.

From March 2012 until August 2014, Peter Hölz successfully completed an internship as student trainee in Engine Development at Porsche Motorsport LMP1, Weissach, Germany. He completed his master studies in Theoretical Mechanical Engineering at KIT in 2015 with the final grade 1.1. The topic of the excellent master thesis was "Thermal investigation of a radial turbine for exhaust energy recovery – analysis of different heat transfer models" at the Institute for Applied Thermofluidics (IATF, Prof. Dr. Xu Cheng, KIT) in cooperation with Porsche Motorsport LMP1, Weissach. The knowledge of this thesis played a key role in the development of the Porsche exhaust energy recovery system and made the development to a success and, therefore, a part of the LMP1 victories in the FIA World Endurance Challenge.

Since November 2015, Peter Hölz had been external doctoral researcher at the Institute of Engineering Mechanics, KIT, in cooperation with and financed by the Dr. Ing. h.c. F. Porsche AG, Department of Engine Development LMP1, Thermomechanical Machine Fatigue, Weissach, Germany. Additionally, he was engaged in teaching students, e.g., in computer tutorials for finite element courses and advanced courses on strength of materials. All his tutorials were evaluated with excellent marks by the students reflecting his attitude on discussing and explaining fundamentals of mechanics. He also supervised two bachelor theses during this time.

As representative of Dr. Ing. h.c. F. Porsche AG in the project committee of the research project "Simulation of Damage Characteristics" of the Research Association for Combustion Engines, Peter Hölz has enriched the project with his own contributions and his almost insatiable interest in computational mechanics. Peter Hölz was highly appreciated by all participants in the research project for his open-mindedness and his authentic appearance.

The following outstanding journal articles which Peter Hölz shared with the scientific community, display and present probably only a small part of his wonderful and great passion he felt by continuously pursuing his future professional dreams:

- Hölz, P. Böhlke, T., Krämer, T.: *Transient temperature calculation method for complex fluid-solid heat transfer problems with scattering boundary conditions.* Applied Thermal Engineering, 149, 1463-1475 (2019)
- Hölz, P., Böhlke, T., Krämer, T.: *Determining water mass flow control strategies for a turbocharged SI engine using a two-stage calculation method.* Applied Thermal Engineering 146, 386-395 (2019)
- Hölz, P., Chebli, E., Böhlke, T.: *Performance enhancement potential of a racing engine with ERS through optimized thermal management.* Proceeding 10th Anniversary THIESEL Conference (2018)
- Hölz, P., Böhlke, T., Krämer, T.: *Fast algorithms for generating thermal boundary conditions in combustion chambers.* Applied Thermal Engineering 141, 101-113 (2018)
- Hölz, P., Böhlke, T., Krämer, T.: *CFD-CHT calculation method using Buckingham Pi-Theorem for complex fluid-solid heat transfer problems with scattering boundary conditions.* Automot. Engine Technol., 1-16 (2018)

The disputation of his excellent doctoral thesis "A dynamic and statistical analysis of the temperature- and fatigue behavior of a race power unit – The effect of different thermodynamic states" was courageously presented on 11th September, 2019 and evaluated with "summa cum laude" by all referees.

Peter Hölz was one of the most mature scientific personalities we have ever met during our scientific career. He acquired his knowledge during his studies in such a breadth and depth which is seldom seen and only rarely the case with exceptionally talented personalities.

In his scientific work, Peter Hölz has made an enormous contribution to a more extensive methodological understanding of the thermodynamic processes in high-performance engines. This contribution was only possible due to the fact that he was able to combine theoretical, numerical and experimental methods in an almost unique way.

Personally, we feel ineffably enriched by having made the acquaintance of Peter Hölz, not only in our inspiring encounters during several scientific discussions, but even more because of his extremely warm, fascinating and passionate personality.

Prof. Dr.-Ing. Thomas Böhlke, KIT, Reviewer of the doctoral thesis

Prof. Dr. sc. techn. Thomas Koch, KIT

Dipl.-Ing. Thomas Krämer, Director Motorsport Quality Management, Dr. Ing. h.c. F. Porsche AG

Prof. Dr. rer. nat. habil. Ulrich Maas, KIT

Prof. Dr.-Ing. Federico Millo, Politecnico di Torino, Italy, Reviewer of the doctoral thesis

Prof. Dr.-Ing. Thomas Seifert, Offenburg University, Reviewer of the doctoral thesis

Zusammenfassung

Rennmotoren können bezüglich hocheffizienter Verbrennung Vorstufen der Serienentwicklung darstellen. In dieser Arbeit wird das dynamische Verhalten eines mager betriebenen Rennmotors hinsichtlich Bauteiltemperaturen, Wärmeflüsse und Kolbenschädigung unter hochzyklischer Last untersucht. Der Fokus liegt dabei auf dem Einfluss verschiedener Motoreinstellungen wie Zündzeitpunkt, Luft-Kraftstoffverhältnis, Kraftstoffmassenstrom, Eingangslufttemperatur und dem Zündsystem wie Standard-Zündkerze und Zündstrahl-Einheit.

Bei der neu vorgestellten Methodik zur stationären und dynamischen Temperaturberechnung werden zur Bestimmung der Wärmeübergangszahl und der Referenztemperatur Druckverläufe gemessen und mittels Wahrscheinlichkeitsdichtefunktionen statistisch verwertet. Die Interpretation als Zufallsvariable ermöglicht eine Separation der Zeitskalen und eine Tiefpassfilterung. Die Methodik beschreibt den Schlepp- und den befeuerten Motorbetrieb, sowie den dynamischen Wärmeübergang innerhalb des Wassermantels. Durch die Transformation der Zufallsvariablen können Abweichungen von den gemessenen Druckverläufen bei nicht allzu inhomogenen Druckänderungen effektiv mittels eines selbst entwickelten Berechnungsmodells behandelt werden. Dabei werden nur zyklus-integrierte Größen benötigt.

Zur Modellreduktion dient die Dimensionsanalyse. Dementsprechend werden zur notwendigen Modellierung des Wärmeübergangs bei verschiedenen Motoreinstellungen die vereinfachten Ansätze nach Bargende, Woschni, Eichelberg und Hohenberg untersucht. Mithilfe des SST (Shear Stress Transport) k-ω Turbulenzmodell von Menter wird ein

Separationsansatz für den Wärmeübergangskoeffizienten innerhalb des Wassermantels vorgestellt. Dabei wird die Orts- und Zeitabhängigkeit separat beschrieben. Zur Validierung dient ein Benzinrennmotor, der mit mehreren Thermoelementen nahe des Brennraums ausgestattet ist. Das thermische Verhalten der verschiedenen Messpositionen kann im stationären und dynamischen Fall simulativ gut abgebildet werden. Bei Verwendung des richtigen Modells können außerdem alle untersuchten thermodynamischen Variationen richtig nachgebildet werden. Der Erwartungswert des Fehlers liegt dabei immer unterhalb von einem Kelvin. Variationen des Luft-Kraftstoffverhältnisses zeigen dabei die größten Temperaturunterschiede.

Zusammen mit einem vereinfachten Einmassenmodell wird die neuartige Berechnungsmethodik zur Untersuchung möglicher Regulierungen von Motortemperaturen und Wärmeflüsse durch alleiniges Kontrollieren des Wassermassenstroms verwendet. Es wird gezeigt, dass mithilfe optimierter Steuer- und Regelungsstrategien dies innerhalb eines großen Bereichs möglich ist. Ausserdem wird durch die Kombination der simulierten dynamischen Temperaturfelder und einer statistischen Zylinderdruckauswertung der Einfluss von verschiedenen Motoreinstellungen auf die Lebensdauer von Kolben untersucht. Dazu werden Schädigungskennfelder, welche auf einem Rissinitiierungsmodell basieren, für den dynamischen Betrieb erstellt. Dabei werden insbesondere unterschiedliche Kombinationen von Gasdruck, Seiten- und Trägheitskraft verglichen. Variationen des Luft-Kraftstoffverhältnisses zeigen wiederum den größten Einfluss auf die Ermüdung.

Die übergeordnete Forschungsfrage dieser Arbeit kann wie folgt formuliert werden: Wie kann eine Methode zur Berechnung des Temperatur- und Ermüdungsverhaltens eines Turbo-Benzinmotors aussehen, welche ohne detaillierte, zyklus-aufgelöste Simulationen die statistischen Druckschwankungen aufeinanderfolgender Verbrennungszyklen berücksichtigt? Dabei ist der Einfluss der Motoreinstellung unter dynamischen Rennbedingungen von besonderem Interesse.

Summary

With regard to series production, race engines can represent precursor stages in high efficient combustion technologies. In this thesis, the dynamic behavior of a lean combustion race engine is investigated with regard to cycle-averaged component temperatures, heat fluxes and, by way of example, piston failure under high cycle fatigue loading. A special focus lies on the influence of various engine settings, like ignition time, air-fuel ratio, fuel mass flow rate, inlet air temperature and the ignition system, such as spark and pre-chamber jet-ignition.
For the introduced stationary, as well as transient, temperature calculation method, in-cylinder pressure curves are measured and processed statistically by using probability density functions for the heat transfer coefficient and the reference temperature. The interpretation as random variables enables a time scale separation with a low-pass filter function. The method includes coasting and fired engine conditions, as well as the dynamic heat transfer within the water jacket. Using an own developed calculation algorithm, which only takes into account cycle integrated quantities and utilizes the transformation of random variables, any deviations from measured pressure curves can be dealt with effectively if pressure changes are not too uneven.
Model reduction is achieved with dimensional analysis. Therefore, concerning the necessary heat transfer modeling for various engine settings, the simplified approaches according to Bargende, Woschni, Eichelberg, and Hohenberg are investigated. For the heat transfer coefficient within the water jacket, using the SST (Shear Stress Transport) k-ω turbulence model by Menter, a separation approach, which separates

the location from the time dependency, is suggested. For validation purposes, a spark-ignition race engine is equipped with thermocouples at various positions in proximity to the combustion chamber. In case of stationary, as well as dynamic, operating conditions, different thermal behaviors of various measuring positions can be simulated successfully. Choosing the right model, all thermodynamic variations examined are in good agreement with theoretical predictions, and the expectation value of the error is in each case less than one Kelvin. Variations in the air-fuel ratio are accompanied with the largest temperature differences.

Together with a simplified lumped-capacity model, the introduced calculation method is used to investigate feasible regulations of engine temperatures and heat fluxes only by controlling the water mass flow rate. It is shown that, with the help of optimized feed forward and control strategies, this is possible within a wide range. In addition, combining the simulated transient temperatures fields and a statistical in-cylinder pressure analysis, the influence of various engine settings on the lifetime of an engine piston is investigated. Therefore, with the help of a crack initiation model, damage maps for the dynamic operation are created. In particular, various constellations of gas pressure, side- and inertia forces are compared. Again, variations in the air-fuel ratio are accompanied with the largest impact on fatigue.

To sum up, the overall research question of this thesis can be formulated as follows: concerning the temperature and fatigue behavior of a turbocharged, spark-ignition engine, what might a calculation method look like, which takes the cycle-to-cycle variability into account, without detailed, crank angle resolved simulations? Thereby, the main goal is to investigate the influence of various engine settings during dynamic operating racing conditions.

Acknowledgments

Special thanks go to Prof. Thomas Böhlke for the supervision of my doctoral thesis and his receptiveness to external research projects. He taught me the constant questioning of existing methodologies and the openness to non-anticipated questions which can particularly arise in interdisciplinary research.

Furthermore, I would like to thank Prof. Federico Millo for co-advising this work, the substantial discussions on thermodynamics and types of failure in engine applications and his great hospitality during my visit at the Politecnico di Torino. I also want to thank Prof. Thomas Seifert for his engagement as my co-supervisor. Within a project of the research association for combustion engines (FVV), I had the pleasure to follow and discuss his excellent work about damage mechanics, which was a true enrichment during my doctorate.

Particular thanks go to Thomas Krämer, former head of engine development at Porsche Motorsport LMP Team (Design and Calculation), for his support and trust over the last years and to Dirk Füllenbach, my supervisor at Porsche Motorsport, for the professional and extracurricular mentoring. In addition, I would like to thank Stefan Moser, current head of engine development at Porsche Motorsport, for the financial and technical capacities, which were necessary for the comprehensive experimental research.

Further, I highly acknowledge the atmosphere created by my colleagues at Porsche Motorsport. Special thanks go to Lars Ramler, Tobias Möllenhoff and Christian Eifrig for their experimental support, as

well as to the mechanics and test bed operators for their high-quality work. They are true world champions.
I would also like to express my gratitude to my colleagues at the Chair for Continuum Mechanics. Despite the external research project and resulting shortened time on site, they have created a cordial atmosphere with many interesting discussions and coffee breaks.
Last but not least, I would like to thank all the people outside of the scientific scope, namely my parents Elisabeth and Paul, my sister Stefanie, my brother-in-law Matthias and all my friends for their wholehearted encouragement and support.

Weissach, September 2019 Peter Hölz

Contents

Chapter 1

Introduction

1.1 Motivation

Present engine development is focused on reducing emissions and fuel consumption in order to meet target fleet-average emissions (Fontanesi et al., 2011). Different approaches like downsizing and turbocharging, which result in higher gas pressure and temperatures, are recommended (Golloch, 2005). Super-lean combustion is also a strategy to decrease heat loss by reducing cylinder gas temperatures. Lambda values near two are reported in Yokomori (2017). In particular, the transient thermal engine behavior influences aspects like fuel consumption or emissions. An optimized thermo-management is crucial in order to ensure engine durability on the one hand, as well as minimize heat- and friction losses on the other hand. Moreover, the engine temperatures have a direct impact on knock protection and volumetric efficiencies (Pang and Brace, 2004). Therefore, but also with regard to a stabilized and robust combustion under transient conditions, their controllability is of interest. To increase the thermodynamic efficiency with optimized combustion processes, lots of engine parameters are available resulting in a complex interaction and, therefore, in a conflict of objectives. As an example, super lean combustion is a strategy to decrease heat loss by reducing cylinder gas temperatures. This possibility of variation requires a better understanding how the engine settings influence component temperatures.

According to a thermomechanical fatigue analysis, the detection of the most critical regions is necessary for fatigue strength evaluation (Fontanesi et al., 2013). As an example, the design and optimization process of water jackets is challenging, resulting in a high demand for experimental techniques and numerical simulations (Fontanesi and Giacopini, 2013). Additional optimization goals are avoiding uncontrolled glow ignition and knocking phenomena both decreasing thermal efficiency and increasing thermomechanical loads. Different answers to this demand involve experimental as well as simulative methods which can include complete automotive cooling systems (Hörmann et al., 2005). Considering, for example, aluminium alloys, the resistance against fatigue is very sensitive to temperature (Robinson et al., 2003). Different fatigue mechanisms like low cycle fatigue (LCF) or high cycle fatigue (HCF) react differently to temperature changes. Furthermore, other fatigue mechanisms like creeping become more important with increasing solid temperatures. The well-known deformation and fracture mechanism maps (Frost and Ashby, n.d.) can give a good overview in each case. To describe quantitatively crack initiation and propagation, which can be caused by the combination of all possible fatigue mechanisms, calculation methods are required to determine solid temperatures. One big challenge is the scattering nature of combustion processes. According to Golloch (2005), some reasons for the typical scattering nature of SI turbo engines are fluctuations in turbulent flow pattern, local differences in gas composition, or decreased local laminar flame speed because of rich or lean combustion. Using large-eddy simulations, Zhao et al. (2018) found that the main reason for different burn rates is the difference in the mean velocity field around the spark gap during ignition. Variations in equivalence ratio fields seem to be less crucial for the flame development and propagation. Therefore, both effects were numerically decoupled by swapping the chemical composition and the velocity field. However, local flame quenching within turbulent flow can also be one reason (Warnatz et al., 2006). Very late ignition times due to knocking control or complex interactions between dynamic boost pressure and exhaust back pressure, resulting in different residual gas, could be other reasons for the scattering behavior (Golloch, 2005). Therefore, due to the non-linearity of heat transfer and damage, a statistical analysis is necessary.

Technology transfer between motor racing and production promotes technical progress (Walliser, 2015). Latest technologies are often used in motorsport for the first time. Although not every technology moves from the development to the mass production stage, the emerging knowledge and methods are at least suitable for similar development projects. By way of example, a lean burn combustion process, which uses jet-ignition, seems to reach consumption levels in gasoline engines that are comparable with diesel engines (Danisch, 2015). As part of the aforementioned downsizing strategy, together with pressure charging, these measures lead to higher mechanical stresses, and, therefore, are a challenge for mass production. Driven by limitations of the fuel flow during racing events, which is applied by technical regulations of high-end motorsport series, modern racing vehicles follow a clear trend towards efficiency development and increased hybridization of the powertrain: Federation Internationale de l'Automobile - FIA (2018a)and Federation Internationale de l'Automobile - FIA (2018b). Therefore, more research activities should focus on the analysis of the temperature- and fatigue behavior of race power units. However, the annual development of new concepts, together with a continuous improvement, requires a quick evaluation and fast calculation methods. Therefore, detailed, combustion-cycle resolved simulations are not always expedient solutions. However, due to the aforementioned non-linearity of heat transfer and damage, cycle-to-cycle variability experimentally observed in SI engines plays a major role. Particularly with regard to temperature analysis and material fatigue, this conflict of objectives is a challenge for future method development, and is an important part of this thesis.

In addition, due to the fact that these engines are hardly operated in a steady state, a dynamic analysis is necessary. Depending on individual race situations, technical regulations or environmental conditions, these kinds of engines additionally undergo a continuous change in thermodynamic state.

1.2 State of the art

1.2.1 Thermal simulation of internal combustion engines

Depending on specific requirements like simulation time, calculation accuracy or modeling of various subsystems, there exist different simulation techniques. Detailed thermal investigations of smaller, isolated subsystems like pistons (Esfahanian et al., 2006), or simplified simulations of complete engines, can be found in literature (Ghebru, 2013).

Early works addressed the problem of engine heat transfer with dimensional analysis and experimental studies: Eichelberg (1939), Woschni (1967), or Hohenberg (1979). An overview about other, simplified approaches can be found in Finol and Robinson (2006). For simplicity, these works, which are mainly based on the Buckingham Pi-Theorem, assume an algebraic correlation for the in-cylinder heat transfer. Quantities which are easy to measure or calculate, like the gas pressure, or the gas temperature with the help of the ideal gas law, are the input parameters. However, the underlying physics of heat transfer, including the turbulent boundary layer behavior, are not directly modeled. In addition, no differential equations, e.g., no balance equations for physical quantities, are directly solved. Nevertheless, these approaches are widely used, especially for complex problems with more subsystems and interactions. In these cases, additional physical modeling effort is necessary for all system parts (Ghebru, 2013).

Moreover, these fast and pragmatic models are used as references and for comparison purposes. As an example, crank angle resolved heat fluxes for a SI engine with direct injection are measured in homogeneous and stratified operation mode in Huegel et al. (2015), and compared with the aforementioned heat transfer models. In case of comprehensive sensitivity analysis, these fast models are an option for heat transfer modeling. As an example, in Sanli et al. (2008), engine speed and load variations, under fired and motored conditions, are investigated regarding heat transfer in an indirect injection diesel engine.

Based on these more phenomenological results, simple lumped turbulence modeling techniques can show the physical background of heat transfer in more detail: using a global k-ε model, Schubert et al. (2005) proposed a heat transfer model with a Reynolds-Colburn analogy. A method which couples detailed CFD techniques with a simplified engine working process analysis to ensure the overall heat transfer rate is proposed in Chiodi and Bargende (2001). The working cycle is based on the model according to Bargende (1991). In this approach, a quasi-dimensional turbulence model is used. Concerning the characteristic velocity, some improvements can be found in (Bargende et al., 2011). An early work about lumped turbulence modeling is Borgnakke et al. (1980). In Grill et al. (2006), within the framework of a quasi-dimensional combustion model, the differential equation for turbulence is supplemented with a special squish term, and the initial condition is used for engine specified adjustments. Further effects on the temporal evolution of the turbulent kinetic energy, like fuel injection, tumble or swirl flow, are presented in Grill (2006). The alternative, quasi-dimensional turbulence model in Liu et al. (2000) contains two differential equations for the turbulent kinetic energy and dissipation. Similarly, Lee and Filipi (2011) models a zero-dimensional energy cascade through a coupled, ordinary differential equation system for the mean kinetic energy and the turbulent kinetic energy. Completing this approach, De Bellis et al. (2014) also considers production terms which are related to the flow through the intake and exhaust valves. For the necessary energy transfer rate, different approaches can be found in literature: (De Bellis et al., 2016), (Fogla et al., 2017), or (Bozza et al., 2018). The first one uses a system of differential equations with the total turbulent kinetic energy and the mean flow kinetic energy. The other two add additional equations for the turbulent dissipation rate and the tumble motion. The last one systematically transforms the three-dimensional conservation equations of the RANS k-ε model into a quasi-dimensional turbulence model. It also provides a reduced three equation concept where the dissipation rate is algebraically derived with the help of the integral length scale. In general, all these works are focused on more physical modeling. Although much more effects and dependencies can be simulated, spatial gradients within the combustion chamber, in case of turbulence and heat transfer, cannot be resolved.

Therefore, many works use three-dimensional CFD in-cylinder flow simulations, including combustion and turbulent heat transfer processes: Payri et al. (2005), Berni et al. (2017), or Mohammadi et al. (2008). Of course, the aims are detailed information about local phenomena such as thermal hot spots. Usually, such points are not directly accessible during experiments. In addition, these methods are cost-effective alternatives to experimental works, e.g., in order to derive simplified correlations (Mohammadi and Yaghoubi, 2010). Moreover, various physical phenomena can be examined separately. For this purpose, turbulent combustion has to be simulated: the fundamental theories and models can be found in Warnatz et al. (2006). A comparison of various simulation techniques for SI engines can be read in Kraus (2007). Using the so called G-equation, Toninel et al. (2009) suggested a coupling method with flamelet libraries for simulating premixed combustion engines. In this context, different simulation approaches regarding turbulent flows, e.g., direct simulations, LES (Large Eddy Simulation) or RANS methods, are discussed in Rodi (2017), or Pope (2010). Various RANS models, including zero, one and two equation models, as well as stress equation models, are given in Nallasamy (1987), and Hölz (2015). In addition, some specific problems like near-wall treatments, low-Reynolds number modeling or curvature effects are addressed, and various applications for different flow structures are given, including numerical methods. With respect to thermal problems, turbulent heat transfer can be described with the help of the turbulent Prandtl number. In Hasan (2007) various models are summarized and compared regarding near wall temperature profiles or Nusselt number correlations. Concerning thermal boundary conditions for internal combustion engines, there are further problems of that sort. Turbulent heat transfer within the water jacket plays an essential role. Conjugate heat transfer methods (CHT) for engines can be found in Fontanesi et al. (2013), or Fontanesi et al. (2011). The special problem of boiling is well reviewed in Warrier and Dhir (2006), Kunkelmann (2011), or Baehr and Stephan (2010a). In this context, some applications to combustion engine water jackets can be found in Fontanesi and Giacopini (2013), Punekar and Das (2013), Mohammadi and Yaghoubi (2011), and Oehrby (2014). In addition, specific models were developed for different kinds of engines or flow structures: concerning heat transfer, HCCI (Homogeneous Charge Compression Ignition) engines are investigated in Broekaert et al. (2017), Chang et al. (2004), and Soyhan et al. (2009). For the last reference, one can find a single generic form for some common models, which allows a more systematic comparison. Similarly, hydrogen engines are studied in Michl et al. (2016), Demuynck et al. (2011), or Shudo et al. (2000).

1.2.2 The effect of different thermodynamic states on engine heat fluxes and temperatures

Regarding heat transfer, the dependence on different engine settings, like ignition time, air-fuel ratio or boost pressure, is of great interest. An excellent review about such sensitivities, in motored and fired engine operation, can be found in Broekaert et al. (2016). In this context, a design of experiments method is applied in Cuyper et al. (2016): various engine settings like, ignition time, air-fuel ratio, fuel, or compression ratio, are investigated. In a more fundamental manner, the Polhausen equation $Nu \propto Re^m Pr^n$ in seven different operating regimes is verified. Therefore, via different exponents m and n, the Nusselt number Nu is correlated with the Reynolds number Re and the Prandtl number Pr. Many other works address the dependence on engine settings, including experimental measurements of heat fluxes and the prediction accuracy of different models: LeFeuvre et al. (1969), Gilaber and Pinchon (1988), Wimmer et al. (2000), Sanli et al. (2008), Sanli et al. (2009), and Sharief et al. (2008). As an example, Wang et al. (2007) investigated heat fluxes as a function of ignition time, air-fuel ratio, and mixture preparation. For instance, the knowledge of the heat transfer is needed for detailed combustion analysis (Wolgast, 2014).

Nevertheless, the questions that arise in this context relate to resulting component temperatures. Most of the previous works investigate crank angle resolved temperature curves: Enomoto and Furuhama (1985), Enomoto and Furuhama (1986b), Enomoto and Furuhama (1989), or Luo et al. (2015). The influence of different engine settings on cycle averaged component temperatures is not studied there. In this context, only highly unique components like nozzle tips are sufficiently represented in literature: Anantharaman and Baskaran (2015), Königsson et al. (2012), or Malbec et al. (2011). It is rare to find publications like Sugihara et al. (2007) where averaged cylinder liner temperatures of a diesel engine are experimentally investigated as a function of ignition time and engine speed. Typically, it can be observed that the cycle averaged mean temperatures react more sensitively to engine settings, in contrast to the corresponding temperature swings. In case of the mentioned cylinder liner measurements, at the thrust side, the temperature differences of the mean value are between approximately three and five times larger than for the temperature waves. In Tillock and Martin (1996), the cylinder head mean temperature is experimentally investigated with regard to three different air-fuel ratios. Again, the influence on the temperature wave form is quite negligible. As another example, Choi et al. (1997) investigates the instantaneous and averaged temperatures at various positions in the cylinder head, the valves, and in the piston of a SI engine. At the exhaust valve bridge, the effect of the engine load on the temperature swing is approximately one order of magnitude lower than on the averaged value. In addition, with increasing speed, the amplitude decreases. Such high frequency temperature waves are also measured in Demuynck et al. (2011) for a hydrogen- and methane-fueled SI engine. In addition, the compression ratio, ignition time, and the mixture richness are varied. Various temperature oscillations on a piston and cylinder head under homogenous and stratified operating conditions can be found in Huegel et al. (2015). In addition, the averaged temperatures are investigated as a function of ignition time. For stratified conditions, the amplitudes on the piston seem to be a little bit larger than in the homogenous case. Further variations, like ignition time, engine load, and engine speed, are measured in Michels et al. (2012). Again, in case of the cylinder head, the temperature waves react less sensitively compared to the averaged values. In addition, in terms of the surface distance, the exponential attenuation of the wave is measured. In this context, an analytical solution for fixed excitation frequencies can be found in Baehr and Stephan (2010b).

1.2.3 Engine temperatures under dynamic operating conditions

Unfortunately, steady-state engine temperatures are significantly different from transient conditions. Within the framework of thermo-mechanical fatigue analysis, many transient FEM simulations can be found: Su et al. (2002), Seifert and Riedel (2009), or Nicouleau et al. (2002).

However, such typical time sections of these self-contained temperature cycles are in the range of one minute, e.g., 0.02 Hz: run-stop conditions should be simulated. Such two- and three-dimensional, transient finite-element models are presented in Rakopoulos and Mavropoulos (1996). It becomes clear that the correct determination of thermal boundary conditions is mandatory. The essential role of engine operational transients, e.g., sudden changes in speed and/or load, is presented in detail. Correspondingly, transient piston temperature measurements, including load and engine speed variations, are reported in Mancaruso et al. (2018). In addition, there also exist combustion-cycle resolved thermal analysis with detailed in-cylinder heat transfer treatment: Mohammadi et al. (2008), Esfahanian et al. (2006), or Kenningley and Morgenstern (2012). However, these simulations describe a high-frequency timescale in the range of the combustion period, e.g, 8-80 Hz, depending on the engine speed. Crank angle resolved calculations correspondingly deal with frequencies of up to several tens of Kilohertz.
In this context, big challenges are transient boundary conditions under transient drives with frequencies between the aforementioned ones, e.g., in the range of 0.1-1 Hz. Continuously changing engine states like transitions between fired and motored condition, including speed build-ups with gear changes and part load sections, are not well presented in literature.

1.2.4 Fatigue analysis for engine applications

Internal combustion engines are subjected to various cyclic mechanical and thermal loads. A general review about fatigue life prediction methods for metals, including crack initiation, crack growth, stochastic methods, damage accumulation, multiaxial loading, energy-based theories, and continuum damage mechanics, can be found in Santecchia et al. (2016). Similarly, high-cycle fatigue criteria, such as critical plane concepts, or theories based on stress invariants, and on the mesoscopic scale approach, are discussed in Papadopoulos et al. (1997). In a more fundamental way, the three basic aspects of metal fatigue, namely cyclic deformation, crack initiation and propagation, is presented in Peralta and Laird (2014).

For engine applications, common challenges are cold starts with subsequent cooling periods, resulting in thermomechanical fatigue. In relation to these macroscopic, inelastic strains, Seifert and Riedel (2010) developed a mechanism-based fatigue procedure for cast iron. With regard to its porosity, which is caused by graphite inclusions, a cyclic plasticity model and a crack growth law, based on the cyclic crack-tip opening displacement, are investigated. Due to high strain amplitudes and the characteristic damage mechanisms for cast iron, crack initiation is neglected. In combination with a parameter identification strategy under isothermal conditions, the proposed model is validated for non-isothermal load cases in Seifert et al. (2010). In particular, for the three different cast iron materials EN-GJS-700, EN-GJV-450 and EN-GJL-250, several complex low cycle fatigue tests, together with conventional tension-compression tests, are conducted. Concerning crack growth modeling under combined low and high cycle fatigue loading, an extension of the method is described in Schweizer et al. (2011). Therefore, using a summation of the two contributions, a ten percent chromium steel is experimentally investigated under room temperature and used for validation purposes. Under additional high cycle fatigue loading, an accelerated crack growth can be observed as soon as a certain crack length is reached. The transition between various crack growth stages remains challenging. An extension for combined thermomechanical and high cycle fatigue loading is presented in Metzger and Seifert (2012). Two more cast iron alloys with a ferritic and an austenitic crystal structure are experimentally investigated in Fedelich et al. (2017). In accordance with Schweizer et al. (2011), the high cycle fatigue frequency, which has been tested up to 20 Hz, has a limited impact on the lifetime. More details concerning the underlying fracture mechanics, like the fatigue crack closure phenomenon and its modeling, can be found in Fischer et al. (2015). A review of necessary modeling approaches for the inelastic material behavior, e.g., plasticity and viscoplasticity, is shown in Chaboche (2008) and Chaboche (1989). Using a notched specimen, representing the inlet and outlet bore holes of cylinder heads, the model of Seifert et al. (2010) is validated for the cast iron EN-GJS-700. Similarly to the aforementioned works, crack growth is the lifetime limiting factor. An application of the method for a V8 Diesel engine is presented in Metzger et al. (2014). With regard to lifetime predictions, several uncertainties are observed: material scatter, temperature errors, and the assumption of small cracks in a homogeneous stress field. Especially the last one causes the conservatism of the model at valve bridge areas. Under engine-like conditions, the thermomechanical fatigue for several cast aluminium alloys is experimentally investigated in Grieb et al. (2010). T6 and T7 conditions are tested and compared with state of the art damage models. In this case, typically for aluminium alloys, the focus is set on crack initiation. In addition, leading to an improvement of the damage prediction, the aging influence is modeled. In Chen et al. (2017), a fatigue analysis for engine cylinder heads, including residual stresses due to previous heat treatment and machining, as well as assembly, procedures, is presented. In addition, concerning cylinder peak pressure, component temperature, and material property variations, a design of experiments is shown. Similarly, thermomechanical fatigue sensitivities in cylinder heads are investigated in Gocmez and Pischinger (2010). Using finite element analysis, an extensive design of experiments method is used for the evaluation of various parameters, like thermal and mechanical loads, as well as the geometry and material. In particular, for the valve bridge, its width is the dominant geometrical factor. Concerning material properties, the thermal conductivity and expansion are the most

significant ones. In summary, the temperature distribution, especially the vertical temperature gradients, determines essentially the damage state.
An application of the aforementioned thermomechanical and high cycle fatigue modeling to a light vehicle diesel AlSiCuNiMg pistons can be found in Kenningley and Morgenstern (2012). In addition to the above-mentioned priorities, the highly transient temperature waves, which are caused by the combustion process and penetrate into the upper piston layers, are discussed in terms of stress contribution. In Myers (1990), the additional creep damage effect on the high and low cycle fatigue performance of an aluminium-silicon alloy, which is often used for engine pistons, is investigated. Therefore, a relatively high temperature, clearly higher than 60 % of the melting temperature in Kelvin, is chosen. Particularly at high cycle fatigue conditions, the effect of creep gets more evident. Similarly, the cyclic deformation behavior of an Al-Si piston alloy with increased amounts of copper and nickel is experimentally investigated in Zhang et al. (2012). Depending on the temperature, hardening as well as softening effects can be observed. Concerning thermal fatigue resistance, further Al-Si alloys are compared with Al2618-T6, which is used in this thesis, in Wunderlich and Hayashi (2012). A good overview about piston fatigue phenomena, including various proposals and solutions, can be found in Dileep et al. (2016). In addition to possible damage cases, characteristic stress fields are discussed in Silva (2006). A piston fatigue evaluation for bi-fuel engines, including engine operations under gasoline and CNG (compressed natural gas) fuel, is presented in Shariyat et al. (2016). In addition, a summary of all relevant factors contributing to fatigue is shown. By way of example, a more material conforming design and fatigue analysis for carbon pistons under stationary engine conditions can be found in Bäumel (2001). In addition, characteristic mechanical and thermal loads, including some FEM modeling details, are presented.

1.2.5 Thermal Management

To increase the engine efficiency by reducing unnecessary heat and friction losses, and decrease its emissions by effective exhaust gas treatments under a transient drive, lots of innovative cooling concepts can be found in literature. A review about engine cooling technologies can be found in Pang and Brace (2004). Different aspects, partially countervailing effects, are discussed. In addition to the already mentioned points, engine protection modes, e.g., improved knock protection, as well as higher volumetric efficiencies for low temperature set points, are mentioned. As an example, in the work of Mohamed (2016) the effect of active engine thermal management on a bi-fuel engine experimentally investigated. A positive effect on heat release rates, fuel consumption and emissions can be proved.

Classical thermal management strategies for combustion engines try to control the water inlet temperature by installing a thermostat, which can direct a part of the water mass flow around the radiator. In Yang (1996), this concept is supplemented with a pump throttle in order to additionally control the heat transfer coefficients. Analogously, by using a reduced order multiple node lumped parameter resistor-capacitor thermal model, Wagner et al. (2003) investigates a variable speed electric pump. In Setlur et al. (2003), a non-linear controller for such applications, additionally adjusting the radiator fan speed, is proposed for cases with unmeasurable heat inputs from the combustion process. A global, asymptotic regulation of the engine temperature can be shown. Similarly, a robust controller design with an estimate of future disturbances is presented in Karnik et al. (2015). Using a zero-dimensional model, Pizzonia et al. (2016) presents a mathematically well formulated controller design procedure for the coolant flow rate, including a discretization of the engine map and its piecewise linearization.

For these kinds of cooling system, Haghighat et al. (2018) reports a lower fuel consumption of 1.1 % under NEDC (New European Driving Cycle)

cycle operation conditions, as well as a reduction of emissions in the range of 5 % for hydrocarbon, and 6 % for carbon monoxide. In addition, a simplified, holistic vehicle model is proposed, which is mainly based on experimental measurements. Similarly, Banjac et al. (2014) developes a transient vehicle model with diverse submodels and corresponding interactions. Again, the application of an electrical water pump results in a fuel saving of about 0.75 % to 1.1 %. In addition, an increase of the turbine outlet temperature can be observed, and hence, a faster catalyst heat-up. A 1D/3D simulation method for the vehicle integrated thermal management is presented in Lu et al. (2016) and Wang et al. (2016). An effective coupling between 1D submodels of single subsytems with the 3D underhood structure offers suitable boundary conditions and parameters for the simulation.
Concerning effective control strategies, there exist many different approaches and the field of control technology is large. The well-known rules according to Kessler are based on direct modifications of the compensated system in the frequency space: the so called symmetrical optimum is proposed in Kessler (1958), whereas Kessler (1955) eliminates the largest time constant in the system. Another, completely empirical, approach is presented by Ziegler and Nichols (1995): the research results are based on many test cases with varying parameters. Using a frequency domain model of the plant, a new method for the auto-calibration of PI and PID controllers is presented in Voda and Landau (1995). Some advantages with regard to the above-mentioned Ziegler-Nichols approach are reported. Concerning some disadvantages of current design techniques for model predictive control, Kothare et al. (1996) proposes a more robust method to deal explicitly with plant model uncertainties. Using dimensional analysis, Fišer et al. (2015) developes a quite general formulation of synthesis of PID control loops with delay. The focus is additionally set on disturbance rejections. A more specific, robust tuning method for first order systems with delay is presented in Yildirim et al. (2015), supplemented by a consideration of stability.

1.3 Outline of the thesis

Several thermal and mechanical problems for internal combustion engines are treated in this thesis. For the purposes of applied research and development, using CFD-CHT and FEM calculation techniques, reduced simulation methods are developed. Therefore, with respect to increasingly shorter development cycles, special focus is placed on model reduction and simplification. In this context, the influence of different engine settings on the temperature- and fatigue behavior is investigated. Except for Chapter 2, each chapter starts with individual research questions. Particular attention will always be given to a dynamic and statistical analysis. For method validation, a spark-ignition turbocharged race engine, including a single cylinder derivative, is available. Due to the aforementioned trend towards optimized thermal efficiency in modern motorsport, special focus is placed on a lean operation mode. Except for chapter two, six, and seven, the chapters are based on published journal publications.

Chapter 2 aims to give possible modeling approaches within a thermal analysis of SI combustion engines. The focus is mainly on the reconsideration of simplified thermal boundary conditions for gas channels, including valves, as well as combustion chambers. The second main objective of this chapter is the presentation of all relevant equations and boundary conditions describing turbulent heat transfer within RANS simulations. All following chapters are mainly based on these theoretical principles.

In Chapter 3, a simplified calculation method is proposed to calculate stationary, cycle averaged temperatures as a function of different engine settings. In relation to Section 1.2.2, a general calculation method is missing. In particular, according to inverse problems, the identification of suitable thermal boundary conditions which can account for different engine settings is challenging and it is not clear how expedient already existing heat transfer models are. That is the focus of the this chapter.

In preparation for the dynamic simulation method in Chapter 5, one of the goals of Chapter 4 is the fast generation of thermal boundary conditions in combustion chambers under fired engine states. In contrast with many other studies, only cycle integrated quantities like the induced torque are needed as input variables, which means that no crank angle resolved in-cylinder pressure data are required.
In relation to the aforementioned absence of suitable simulation methods within frequency ranges of 0.1-1 Hz in Section 1.2.3, Chapter 5 presents a simplified, transient temperature calculation method. This includes engine speed build-ups under gear changes, torque variations, or the transition from fired to coasting conditions.
Chapter 6 combines the methods of Chapter 3 and 5. Therefore, the investigation of different engine settings, including the identification of suitable thermal boundary conditions, on cycle averaged engine temperatures is extended to transient states.
In Chapter 7, by the example of an engine piston, dynamic race conditions like changing temperature fields, and engine speeds, are investigated numerically concerning fatigue. This accomplishes already existing works which mainly investigate stationary temperature fields or cold starts, as well as cooling phases, with time scales of several seconds. With the help of Chapter 6, and in-cylinder pressure measurements, different engine settings are also investigated.
Concerning control and feedback control systems for engine temperatures, one of the goals of Chapter 8 is the systematic investigation of their controllability under highly transient racing conditions by only adjusting the water mass flow rate within the water jacket, e.g., at a constant water inlet temperature. Therefore, a two-stage calculation method, consisting of the three-dimensional simulation approach of Chapter 5, and a simplified lumped capacity model, is proposed. This complements existing published works which usually control more than one parameter, but use one model.
Finally, chapter 9 gives the main results and an outlook for future works.

Chapter 2

Turbulent heat transfer: A reduced, holistic approach for internal combustion engines

2.1 Outline of the chapter

The chapter presents various methods of model reduction within the framework of CFD-CHT calculations. For the purpose of an holistic thermal engine simulation, individual boundary conditions for the gas, oil and water region are summarized.
In case of the gas channels and the combustion chamber, for reasons of rapidity, statistics and transient simulations in the following sections, the chapter attaches great importance to algebraic and simple, ordinary differential equations, instead of partial differential equations of a three-dimensional combustion simulation with heat transfer processes. Therefore, boundary conditions can easily be obtained from test bench measurements, e.g., in-cylinder gas pressure and flow rates. This also ensures fast comparisons between engine settings. In addition, a brief summary of the Buckingham Pi-Theorem is given. Afterwards, engine specific correlations including modeling approaches for temperature dependent gas properties are given. Regarding the spatial distribution of heat transfer coefficients within the combustion chamber, some modeling techniques such as detailed CFD results and experimental measurements

are given. Piston specific requirements, like oil jet cooling or heat conduction to the cylinder liner, are also presented.
Furthermore, the turbulent heat and fluid flow in engine water jackets is described with the SST k-ω turbulence model by Menter (Menter, 1994). Special focus is placed on turbulent boundary layer modeling including wall laws. Increased heat transfer due to turbulence and its modeling according to a Boussinesq approximation including temperature wall laws is discussed.

2.2 Determination of boundary condition using Buckingham Pi-Theorem

2.2.1 Buckingham Pi-Theorem

In the following, a brief summary about the Buckingham Pi-Theorem is given. (For details, see (Bridgman, 1932)). Given is an algebraic physical relation F between z_d dimensioned quantities Q:

$$F(Q_1, Q_2, ..., Q_{z_d}) = 0. \tag{2.1}$$

For purely mechanical systems, there are the three fundamental quantities mass, length and time. In thermomechanical systems, the temperature is added to the aforementioned quantities. The Pi-Theorem states that, equivalently to Eq. (2.1), there exists a new relation $\tilde{F}$ with $z_d - r$ dimensionless quantities Π of the form:

$$\tilde{F}(\Pi_1, \Pi_2, ..., \Pi_{z_d - r}) = 0, \tag{2.2}$$

with r being the rank of the dimension matrix. The matrix entries are the exponents $\tilde{\alpha}_{i,j}$ which correlates the dimension of a quantity Q_j with a fundamental quantity A_i according to

$$[Q_j] = \prod_i A_i^{\tilde{\alpha}_{i,j}}. \tag{2.3}$$

Assuming that the heat flux vector $\boldsymbol{q}$ can be written as

$$\boldsymbol{q} = \alpha \left(T_{\text{ref}} - T_{\text{s}}\right) \boldsymbol{n}, \tag{2.4}$$

with a heat transfer coefficient α which is independent of the temperature difference between fluid and solid. This is called Newton hypothesis. Note that the heat transfer coefficient α refers to a well-defined reference temperature T_{ref}. In case of engines, examples are characteristic gas temperatures in channel or in-cylinder flows. T_{s} is the solid temperature. $\boldsymbol{n}$ is the normed surface normal vector. Other forms of heat transfer are free convection or boiling. Within turbulent flows, one example for Eq. (2.1) is the following form (Lienhard and Lienhard, 2003):

$$\alpha = f\left(\lambda, l, \rho, c_p, v, \mu\right), \tag{2.5}$$

where l is a characteristic length scale of the considered system and v is a characteristic velocity. Thermal conductivity λ, density ρ, heat capacity at constant pressure c_p and dynamic viscosity μ refer to the fluid. A dimensionless form (2.2) can be derived by using the Buckingham Pi-Theorem:

$$Nu = \tilde{f}\left(Re, Pr\right), \tag{2.6}$$

with $Re = lv\rho/\mu$, $Pr = \nu/a$ and $Nu = \alpha l/\lambda$. $a = \lambda/\rho c_p$ is the temperature conductivity and $\nu = \mu/\rho$ the kinematic viscosity.

2.2.2 Modeling temperature-dependency of gas mixture properties

Within engine simulations, Eq. (2.5) can be used for inlet and exhaust channels as well as in-cylinder heat transfer. Therefore, the temperature-dependency of the gas mixture properties λ, μ and c_p has to be considered. One possible method is to model the properties of individual components with a subsequent homogenisation. The first modeling can be done by using measured data for the heat capacity at constant pressure c_p as well as model equations which are based on the kinetic theory of ideal gases with rigid spheres and the subsequent modeling of real gas effects with the help of reduced collision integrals. Temperature-dependent heat capacities are modeled by using the so called *NASA polynomials*. Detailed measurement data for many chemical species can be found in McBride et al. (1993). The heat capacity c_p is modeled with two polynomials of degree four for two different temperature ranges with the transition temperature of 1000 K. Under the assumption of ideal gases, the heat capacity at constant volume c_v can be calculated with the help of the universal gas constant R. The transition between the two polynomials is continuous and continuously differentiable. Actually, the *NASA polynomials* are related to the amount of substance and not to mass. The heat capacities of the mixtures are gained from averaging with respect to the mole fractions. For the thermal conductivity λ of a species, following model can be used (Warnatz et al., 2006):

$$\lambda = \frac{25}{32} \frac{\sqrt{\pi m_p \tilde{k} T_{\text{ref}}}}{\pi \sigma_{\text{Pa}}^2} \frac{c_v}{\Omega^{(2,2)*} m_p}, \tag{2.7}$$

with the particle mass m_p and its diameter σ_{Pa}. $\tilde{k}$ is the Boltzmann constant. For once, c_v is the molecular heat capacity in Eq. (2.7). For the dynamic viscosity, following relation can be used (Warnatz et al., 2006):

$$\mu = \frac{5}{16} \frac{\sqrt{\pi m_p \tilde{k} T_{\text{ref}}}}{\pi \sigma_{\text{Pa}}^2} \frac{1}{\Omega^{(2,2)*}}. \tag{2.8}$$

The reduced collision integral $\Omega^{(2,2)*}$ takes real gas effects into account. Under the assumption of a Lennard-Jones-6-12-potential, this integral is a unique function of the reduced temperature $T^* = \tilde{k}T_{\text{ref}}/\tilde{\epsilon}$ where $\tilde{\epsilon}$ is the depth of the intermolecular potential. Values for σ_{Pa} and $\tilde{\epsilon}$, as well as details to the aforementioned models and the reduced collision integral, can be found in Warnatz et al. (2006). To approximate the properties of the gas mixture, following empirical correlations are suggested (See, e.g., Mathur et al. (1967)):

$$2\lambda = \sum_i x_i \lambda_i + \left(\sum_i \frac{x_i}{\lambda_i} \right)^{-1}, \tag{2.9}$$

$$2\mu = \sum_i x_i \mu_i + \left(\sum_i \frac{x_i}{\mu_i} \right)^{-1}. \tag{2.10}$$

λ_i and μ_i are the properties for species i and can be calculated according to Eq. (2.7) and (2.8). x_i is the corresponding mole fraction. Equation (2.9) and (2.10) are combinations of the arithmetic and harmonic mean. In this so called Voigt-Reuss-Hill average, the corresponding quantities, which are comparable with stiffnesses and compliances, are taken into account to the same extent.

2.2.3 Heat transfer modeling in the inlet and outlet system

The correlation according to Caton and Heywood (1981)

$$Nu_{\mathrm{v}} = C_1\ Re_{\mathrm{v}}^{0.58}(D_{\mathrm{v}}/l_{\mathrm{v}})^{0.2} \tag{2.11}$$

is one possible approach for the outlet valve stem and its back surface. The constant C_1 in Eq. (2.11) depends on many factors like exact geometry and its tolerances. Values about 0.92 to 1.84 are reported in Caton and Heywood (1981). The dimensionless numbers Re_{v} and Nu_{v} are the Reynolds number and Nusselt number which are based on the valve lift l_{v} and the exhaust jet gas velocity. D_{v} is the diameter of the valve. In Caton and Heywood (1981), different modeling approaches are investigated according to dominant heat-transfer modes in an exhaust port. Experimental results for exhaust gas temperatures are compared with computed solutions which are based on a finite difference calulcation algorithm for the exhaust port. Equation (2.11) is a modified result from a two-dimensional jet impingement model. Because of the outlet valve stem and the curved shape of the exhaust port various large-scaled eddies occur. Assuming the eddy turn-over time less than the eddy's port residence time, following large-scale model is proposed for the exhaust port:

$$Nu = C_2\ \sqrt{Re_{\mathrm{j}} Pr}. \tag{2.12}$$

Again, the constant C_2 depends on the exact configuration. Values in the range of 1.60 to 1.92 are reported in Caton and Heywood (1981). The dimensionless numbers Nu and Re_{j} are based on the duct diameter and the exhaust jet velocity through the valve opening v_{j}. According to Heywood (1988), such jet conditions are approximately fulfilled if $l_{\mathrm{v}}/D_{\mathrm{v}} < 0.2$ holds. Fig. 2.1 shows a comparison between Eq. (2.12) and a conventional approach according to $Nu \propto Re^{0.8}$, which is a common correlation for turbulent pipe flow. As can be seen, the large-scale model shows better agreement with the measured temperature profiles as well as better agreement for the average temperature.

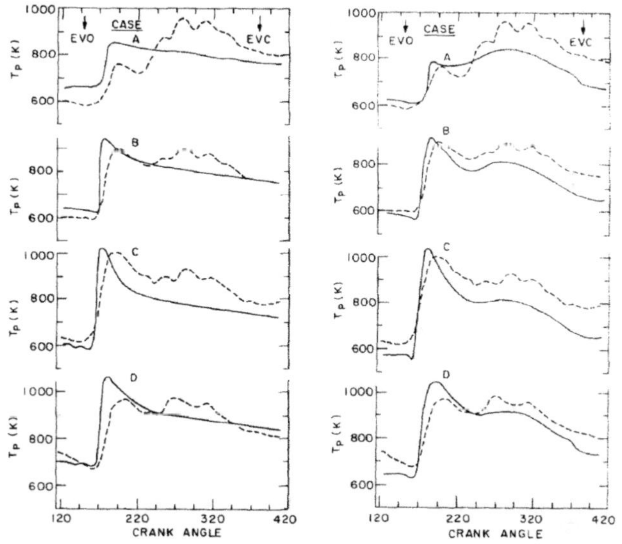

Figure 2.1: Computed and measured (dashed line) port exit gas temperature T_p for four different engine conditions A-D during exhaust valve opened (EVO) and exhaust valve closed (EVC) time interval. Comparison between the large-scale model (right) and a conventional approach according to $Nu \propto Re^{0.8}$ (left). Figure reprinted from Caton and Heywood (1981) (with permission from Elsevier **https://doi.org/10.1016/0017-9310(81)90003-X**).

For more straight sections of exhaust pipes, a conventional turbulent pipe flow can be assumed with an exponent of $m = 0.78 - 0.8$ in Eq. (2.15) (See, e.g., Heywood (1988) and Caton and Heywood (1981)). Another form for the exhaust port is suggested in Heywood (1988): $Nu \propto \overline{Re}^{0.8}$. The special Reynolds number $\overline{Re}$ is based on the time-averaged gas velocity. For the inlet valve stem, an expression similar to Eq. (2.11) can be assumed (Yang et al., 2000). Therefore, the coefficient has to be reduced by 40 %. Different engine conditions like variations in air-fuel ratio, ignition time or engine speed are investigated with respect to inlet and outlet valves in Yang et al. (2000). Compared with measurements, the proposed models show convincing results. The heat transfer coefficients are based on the inlet and exhaust gas temperature. Therefore, a great dependence on the exhaust temperature can be observed. Additional approaches for the back surface of inlet valves can be found in Wang and Berry (1985). Following correlation is suggested for closed valves describing natural convection of the back surface:

$$Nu = C_3 \, (Pr \, Gr)^{0.25} \, . \tag{2.13}$$

The constant C_3 in Eq. (2.13) mainly depends on the product of the Grashof number Gr and the Prandtl number Pr. The first one is defined by

$$Gr = \frac{g\tilde{\beta}\Delta T l^3}{\nu^2}, \tag{2.14}$$

with gravity g, a characteristic temperature difference ΔT and the coefficient of thermal expansion $\tilde{\beta}$. It describes the ratio between buoyancy forces due to temperature differences and viscous forces, which are multiplied with the Reynolds number. For $10^4 \leq Pr \, Gr \leq 10^9$, a typical value is 0.54. For larger products, the constant is in the range of 0.14 (See, e.g., Wang and Berry (1985)). The characteristic length scale of Eq. (2.13) is the valve diameter, and the characteristic temperature difference is defined with solid and fluid temperatures. In literature, typical dimensionless forms of the type (2.6) has been proposed for pipe flows: Boelter et al. (1965), Colburn (1964), Gnielinski (1976) or Petukhov (1970). The last two references include the effect of surface roughness. For non-circular ducts, the hydraulic diameter is often used (Lienhard and Lienhard, 2003). Regarding the intake port of engines, lots of experimental work was done by Uchida et al. (2005), who investigated the heat transfer in intake ports under various conditions. Neglecting the Prandtl number influence, a simple form

$$Nu = cRe^m, \tag{2.15}$$

with a coefficient c and an exponent m is suggested. The experiments showed that the coefficient gets smaller for longer ducts, that means for more developed flow. In contrast, the exponent m gets slightly larger. This is in accordance with other observations (Sellars et al., 1954). Equation (2.16) and (2.17) are empirical suggestions for the coefficient c and exponent m, which are derived from experiments in Uchida et al. (2005).

$$c = 0.02\left(1 + 1.68\exp -0.0018\left(\frac{l_s + l_p}{D_{\text{hyd}}}\right)^2\right), \tag{2.16}$$

$$m = 0.8\left(0.0041\left(\frac{l_s + l_p}{D_{\text{hyd}}}\right) + 1.22\right). \tag{2.17}$$

The characteristic length scales l_s, l_p and D_{hyd} are the length of the straight pipe, the curved intake port and the hydraulic diameter, respectively. Detailed investigations about spatially resolved expressions or curved geometries can be found in Bauer et al. (1998). As an example, the local Nusselt number of the entrance region of a straight duct is shown in Fig. 2.2. Heat transfer increase due to the thermally underdeveloped boundary layer results in higher Nusselt numbers at the entrance. Two model equations for straight and curved ducts are given in Bauer et al. (1998):

$$Nu_x = aRe_x^b\left(1 + c\exp\left(d\frac{x}{D_{\text{hyd}}}\right)\right), \tag{2.18}$$

$$Nu = eRe^b. \tag{2.19}$$

In this case, Nu_x and Re_x are local quantities due to changes in gas properties and velocities with the axial coordinate x. In Eq. (2.19), spatial averaged values are used. Corresponding model parameters for Eq. (2.18) and (2.19) can be found in Table 2.1. Compared with straight pipes, secondary flow patterns in curved pipes can increase heat transfer.

Manifold	**a**	**b**	**c**	**d**	**e**
Straight	0.043	0.73	2.9	-0.68	0.062
Curved	0.13	0.66	0.69	-1.05	0.14

Table 2.1: Model parameters in Eq. (2.18) and (2.19). Table reprinted from Bauer et al. (1998) (with permission from Elsevier **https://doi.org/10.1016/S0142-727X(97)10001-7**).

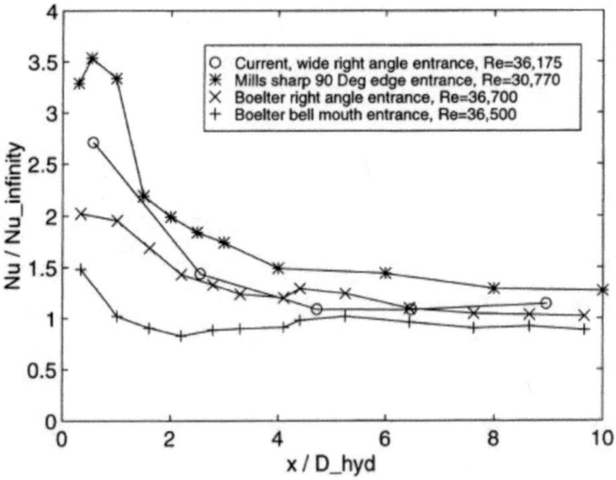

Figure 2.2: Local Nusselt number divided by fully developed value Nu_{infinity}. Reported data for steady flow in straight duct at particular Reynolds number in comparison to literature data. Figure reprinted from Bauer et al. (1998) (with permission from Elsevier **https://doi.org/10.1016/S0142-727X(97)10001-7**).

According to the influence of the thermal entrance region, one has to add a new dimensionless number to Eq. (2.6):

$$Nu = \tilde{f}\left(Re, Pr, x/l\right), \tag{2.20}$$

with a spatial coordinate x and a characteristic length scale l. Equation (2.20) is often summarized with the help of a new dimensionless number, the Graetz number $Gz = Re\ Pr\ l/x$:

$$Nu = \tilde{f}\left(Gz\right). \tag{2.21}$$

2.2.4 Heat transfer modeling for oil flow

Using the Pi-Theorem for laminar flow conditions, there is only one dimensionless group

$$\alpha = f(D_{\text{hyd}}, \lambda). \tag{2.22}$$

In case of a uniform heat flux at the wall, Eq. (2.22) reduces to the very simple form $Nu = 4.364$. For an isothermal pipe, one can show that $Nu = 3.657$ holds (Lienhard and Lienhard, 2003). For pipe sections in which turbulent flow occurs, the formula after Gnielinski (Gnielinski, 1976) is one possible choice:

$$Nu = \frac{(f/8)(Re - 1000)Pr}{1 + 12.7\sqrt{f/8}(Pr^{2/3} - 1)}, \tag{2.23}$$

with

$$f = \frac{1}{(1.82 \log_{10} Re - 1.64)^2}. \tag{2.24}$$

In this case, Nu and Re are based on the pipe diameter. The above formula can be used up to Reynolds numbers $Re \leq 5 \cdot 10^6$ and Prandtl numbers $Pr \leq 2000$ (Lienhard and Lienhard, 2003).

2.2.5 Heat transfer modeling in the combustion chamber

Different strokes of the engine cause different main flow patterns in the cylinder: a tumble flow in the intake and outlet stroke, which is temporally interrupted by a squish flow. Depending on the channel geometry, an additional swirl flow structure can occur. Time-averaged correlations can be found in Annand (1963) or Taylor and Toong (1957). Compared to convective heat transfer, radiation heat transfer in conventional spark-ignition engines is generally small (Heywood, 1988). Assuming quasi steady conditions at every time step, one possible approach for the Buckingham Pi-Theorem (2.1) has the following form (Heywood, 1988):

$$F(\alpha, B, z_c, y_1, y_2, ..., v, u_1, u_2, ..., \lambda, \mu, c_p, \rho, n_{\text{engine}}, \alpha_{\text{cr}}, \dot{q}_{\text{chem}}, T) = 0. \tag{2.25}$$

In this case, the heat transfer coefficient in the combustion chamber α depends on the engine bore B and its axial cylinder length z_c, as well as any number of additional geometrical dimensions y_i. In an analogous manner, this idea can also be used for a characteristic velocity v and different velocity ratios u_i. The engine speed is described by n_{engine}, and α_{cr} describes the crank angle. The material properties in Eq. (2.25) are related to the fluid. A characteristic temperature T is also taken into account. The chemical energy release rate per unit volume is described by $\dot{q}_{\text{chem}}$. Using mass, length, time and temperature as the fundamental quantities, this results in the following dimensionless form (Heywood, 1988):

$$\tilde{F}\left(\frac{\alpha B}{\lambda}, \frac{\rho v B}{\mu}, \frac{c_p \mu}{\lambda}, \frac{c_p T}{v^2}, \frac{n_{\text{engine}} B}{v}, \frac{\dot{q}_{\text{chem}}}{\rho c_p n_{\text{engine}} T}, \frac{z_c}{B}, y_1, y_2, ..., u_1, u_2, .., \alpha_{\text{cr}}\right) = 0. \tag{2.26}$$

The first three dimensionless numbers are the Nusselt, Reynolds and Prandtl number. Remembering the isentropic state change $\partial p / \partial \rho$, describing the square of the sound speed, the fourth number can be interpreted as a Mach number, which is small for engine applications. In most cases, the characteristic velocity v contains a part which is proportional to the mean piston speed, which in turn is proportional to the engine speed n_{engine}. So, the fifth number is nothing else than another geometrical dependence. Remember that Eq. (2.26) describes instantaneous, spatially averaged heat transfer coefficients. For the reference temperature, the cylinder-average gas temperature

$$\overline{T}_g = \frac{pV}{NR} \tag{2.27}$$

is used. In Eq. (2.27), N is the amount of substance, V is the total volume, and p is the static pressure. One has to mention that the very early attempts to model heat transfer coefficients does not use the form (2.26). Instead, they use correlations of the form

$$\alpha = f\left(v_{\text{piston}}, p(t), \overline{T}_g(t)\right). \tag{2.28}$$

The mean piston speed is described by v_{piston}. The physical time is given by t. A very good overview, including experimental results to both attempts (2.26) and (2.28), can be found in Adair et al. (1972).
In addition, there are suggestions for the Reynolds and Prandtl exponents in a simplified form of Eq. (2.26):

$$Nu \propto Re^m Pr^n, \tag{2.29}$$

which is based on correlations for forced convection in turbulent flow inside circular tubes (Colburn, 1964). Table 2.2 shows three popular models for Eq. (2.26) and (2.28). A value of $m = 0.8$ is typical. The model parameters $C_{\text{Eichelberg}}$, C_{Woschni} and $C_{\text{Hohenberg}}$ are calibration values. The diameter of a sphere with the same volume as the instantaneous cylinder

Model	HTC	Characteristic velocity
Eichelberg (1939)	$\alpha = C_{\text{Eichelberg}} v^{1/3} \left(p\overline{T}_g\right)^{1/2}$	$v = v_{\text{piston}}$
Woschni (1967)	$\alpha = C_{\text{Woschni}} B^{m-1} p^m v^m$	$v = c_1 v_{\text{piston}}$
	$\overline{T}_g^{0.75-1.62m}$	$+c_2 V_{\text{d}} \left(\overline{T}_g / (pV)\right)_{\text{IVC}} \Delta p_m$
Hohenberg (1979)	$\alpha = C_{\text{Hohenberg}} d_{\text{s}}^{m-1} p^m v^m$	$v = \left(\overline{T}_g / \text{K}\right)^{-0.1625}$
	$\overline{T}_g^{0.75-1.62m}$	$\left(v_{\text{piston}} + 1.4\ \text{m/s}\right)$

Table 2.2: Heat transfer coefficients in the combustion chamber for three popular models.

volume is described by d_{s}. The pressure difference $\Delta p = p - p_m$ describes the combustion effect. Usually, the unit of the static pressure p is *bar*. The physical quantities V_{d} and p_m are the displaced volume and the motored pressure, respectively. For the model by Woschni, the constant c_2 is zero, except in the combustion and expansion period. A value of $c_2 = 3.24\text{e}^{-3}$ m/(sK) is recommended. For the constant c_1, a value of 6.18 in the gas exchange period, and 2.28 else, is suggested. One can find some modifications to the constants according to velocity increase because of swirl motion (Gamma Technologies, 2016). Details to effects which result from additional swirl motion can be found in Sihling and Woschni (1979). In this thesis, the model according to Bargende is additionally investigated (Bargende, 1991). In the following, only the most important equations for this heat transfer model are given. In complementarity with the original formulation, the characteristic velocity v additionally consists of a scaled combustion convection v_{c} (Bargende et al., 2011):

$$v = \sqrt{(8/3)k + v_\mathrm{p}^2 + v_\mathrm{c}^2}, \tag{2.30}$$

$$v_\mathrm{c} = \sqrt[6]{y_b}\,\frac{B}{4}\left(\frac{\mathrm{d}y_b}{\mathrm{d}t} - \frac{T_\mathrm{ub}}{\overline{T}_g}\frac{\mathrm{d}x}{\mathrm{d}t}\right). \tag{2.31}$$

Using the turbulent kinetic energy k and the current piston speed v_p, the other two summands describe velocity fluctuations due to turbulence and the in-cylinder flow structure. According to a two zone combustion model, the ratio between burnt and complete in-cylinder volume is given by y_b. Similarly, the mass fraction is given by x. Beginning with the ignition time, the unburnt gas temperature T_ub is calculated with the assumption of a polytropic compression and an homogeneous in-cylinder pressure. According to Lejsek et al. (2010), the burnt gas temperature is modeled by using the volume balance of the two zones and the ideal gas law. Together with the pressure indication measurements $p(\alpha_\mathrm{cr})$, the necessary burn function is gained with a pressure trace and combustion analysis. To calculate the turbulent kinetic energy k in Eq. (2.30), a lumped turbulence model is necessary. In this thesis, the model according to Borgnakke et al. (1980) is used. Assuming isotropic, homogenous turbulence for equilibrium conditions, the turbulent dissipation rate ε is given by $\varepsilon \sim k^{3/2}/l$ with a characteristic eddy length scale l, which is given by the combustion chamber volume V according to $l = (6/\pi V)^{1/3}$. On the basis of the rapid distortion theory, e.g., serving the angular momentum, the turbulent kinetic energy k is related to the eddy length scale l according to $k^{1/2} \sim l$. The conservation of mass finally gives the turbulent production rate $\mathrm{d}k \sim 2k/(3\rho)\,\mathrm{d}\rho$, resulting in the following differential equation:

$$\frac{\mathrm{d}k}{\mathrm{d}t} = -\frac{2}{3}\frac{k}{V}\frac{\mathrm{d}V}{\mathrm{d}t} - \varepsilon_\mathrm{c}\frac{k^{3/2}}{l}. \tag{2.32}$$

Together with the initial condition at the inlet valve closed state, the model constant ε_c is aligned with a three-dimensional in-cylinder CFD simulation for a representative engine speed of 7000 rpm. The sensitivity of the initial value to the engine speed is then modeled according to Kozuch (2004). The equation model of Liu et al. (2000), which is mentioned in the introduction, contains a detailed approach for the two proportional constants $\varepsilon \sim k^{3/2}/l$ and $\mathrm{d}k \sim 2k/(3\rho)\,\mathrm{d}\rho$.

2.2.6 Spatial distribution of heat transfer coefficients

Based on the complex interaction between flame arrival time, local gas velocity as well as compressibility effects, a spatial distribution of the HTC within the combustion chamber should be modeled. Therefore, detailed 3D-CFD simulation results can be used. In particular, the relative distribution between different components is of interest. As an example, Fig. 2.3 shows instantaneous heat transfer coefficients for different engine components. Some tests to this topic are conducted in LeFeuvre et al. (1969) or Dent and Sulaiman (1977). In addition, some extensions to common correlations for the heat transfer coefficient with a spatially dependent characteristic velocity and length scale are proposed. Therefore, engines with a pronounced swirl motion within the cylinder are used. However, one has to pay attention to the kind of engine: spark-ignition engines behave differently than compressed-ignition aggregates. Moreover, the reference temperature can be an averaged value or a local variable.

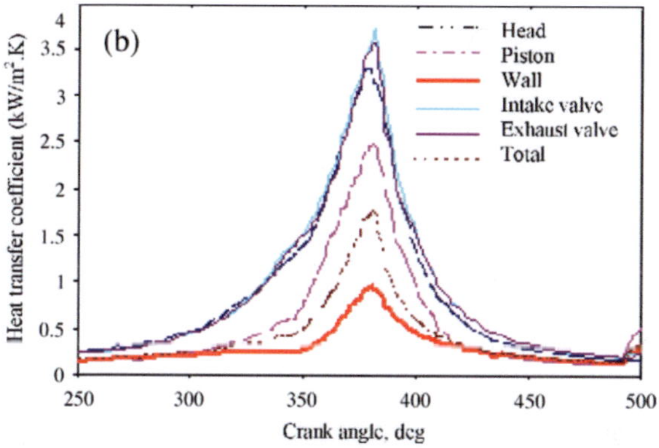

Figure 2.3: Instantaneous heat transfer coefficients for diverse engine components. Detailed CFD results for a 2 liter spark-ignition engine at 3500 rpm are shown. Figure reprinted from Mohammadi et al. (2008) (with permission from Elsevier **https://doi.org/10.1016/j.icheatmasstransfer.2007.06.008**).

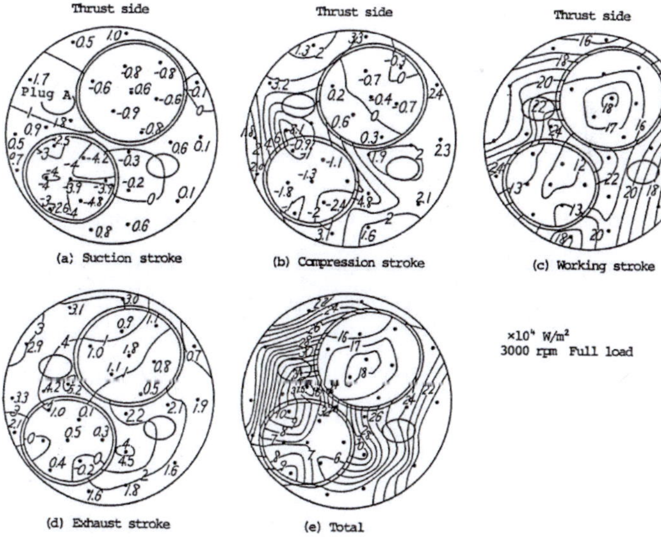

Figure 2.4: Heat flux distribution on the cylinder head and the valves for each stroke. The asymmetrical assembly situations for the spark plug are indicated with an ellipse. Figure reprinted from Enomoto and Furuhama (1986a) (with permission from JSME **https://doi.org/10.1299/jsme1958.29.2196**).

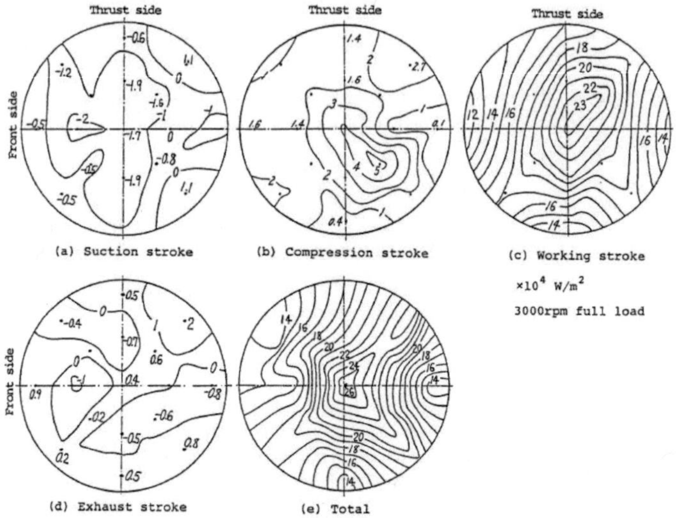

Figure 2.5: Heat flux distribution on the piston crown for each stroke. For the asymmetrical assembly situation for the spark plug, see Fig. 2.4. Figure reprinted from Enomoto et al. (1985) (with permission from JSME **https://doi.org/10.1299/jsme1958.28.647**).

Measurements by Enomoto et al. (1985) on a spark-ignition engine show a heat flux distribution on piston crown, which is approximately linear to the radial position. Minimum 85 % of the total heat flux results from the working cycle. Fig. 2.5 shows isolines of measured heat fluxes on a piston crown. A similar distribution for the cylinder head and valves can be seen in Fig. 2.4. The asymmetrical isolines can be seen much better in case of the cylinder head than the piston crown. However, all these ratios are integrated results over the complete engine cycle. Results for instantaneous heat transfer coefficients in Enomoto and Furuhama (1989) show much complicated ratios as a function of the crank angle. As an example, Fig. 2.6 shows such instantaneous coefficients for different radial measuring positions on the cylinder head. Due to different flame arrival times, one can clearly recognize a phase shift with radial distance. Except for some local hot spots, the intake and exhaust valve surfaces are weakly inhomogeneous. Homogeneous boundary conditions within CHT simulations can also be found in Fontanesi et al. (2013). In addition, a parabolic distribution of the heat flux over the cylinder head, centered at the spark plug, is used. Regarding the cylinder liner, an axial distribution of the HTC should be taken into account. Different approaches for the spatial distribution on engine liners are investigated in Schwiderik (2013) with FEM calculations.

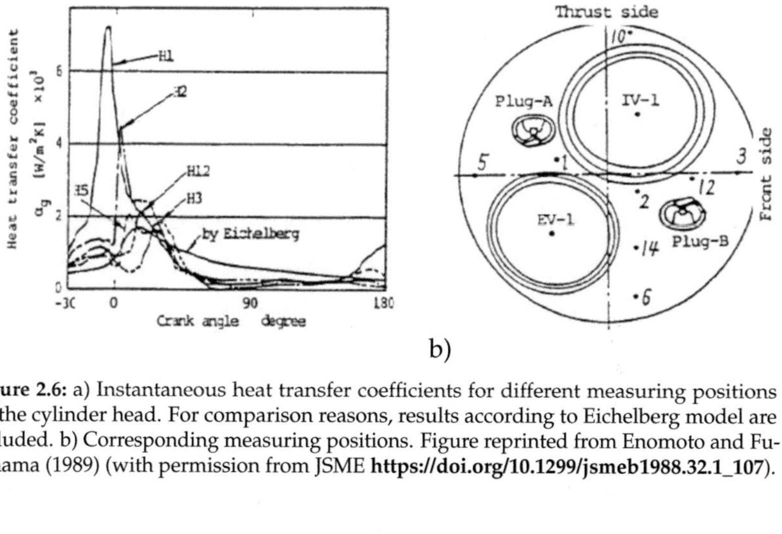

a) b)

Figure 2.6: a) Instantaneous heat transfer coefficients for different measuring positions on the cylinder head. For comparison reasons, results according to Eichelberg model are included. b) Corresponding measuring positions. Figure reprinted from Enomoto and Furuhama (1989) (with permission from JSME **https://doi.org/10.1299/jsmeb1988.32.1_107**).

As an example, Fig. 2.8 and 2.7 show two heat flux distributions for a diesel and spark-ignition engine. In the upper part of the liner, significantly higher heat fluxes can be observed. This can be explained by the direct contact to the propagating flame and a longer contact time with the exhaust gases. However, regarding the axial position, in Enomoto et al. (1985) a more step-wise decrease of the heat flux is measured. A possible explanation for the first drop is the penetration of the combustion gas up to the piston rings (Rakopoulos and Mavropoulos, 1996). In Fig. 2.8, one can also recognize a rapid decrease in the heat flux in the uppermost area of the liner. According to Rakopoulos and Mavropoulos (1996), possible explanations are local flow phenomena with frictional resistance to the gas flow.

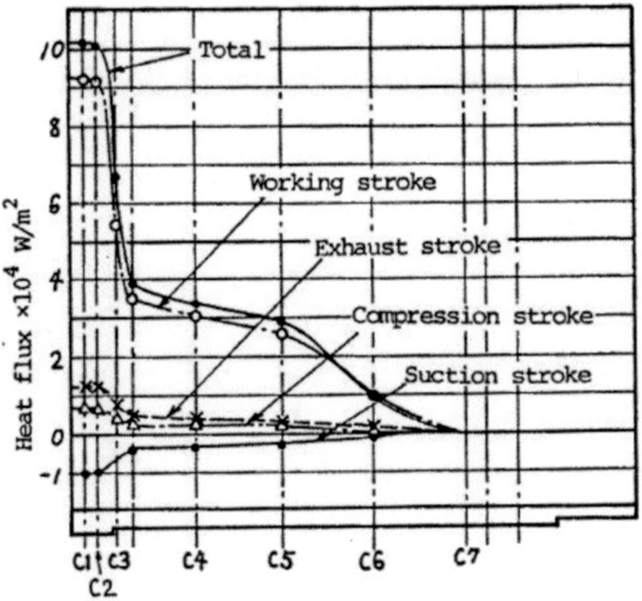

Figure 2.7: Heat flux in the cylinder liner of a spark-ignition engine for different axial distributed measuring positions C1-C7. Figure reprinted from Enomoto et al. (1985) (with permission from JSME **https://doi.org/10.1299/jsme1958.28.647**).

The increasing share of friction and piston heat conduction with growing axial distance is investigated in Woschni (1979). In the mid area, the ratio is about 50 %. Own studies show that the conduction part is larger than the friction part if the engine speed is not too high. However, the lower part of the liner is also wetted with air and oil from the engine crankcase. Heat transfer coefficients are in the range of 100-200 W/m^2K (Rakopoulos and Mavropoulos, 1996).

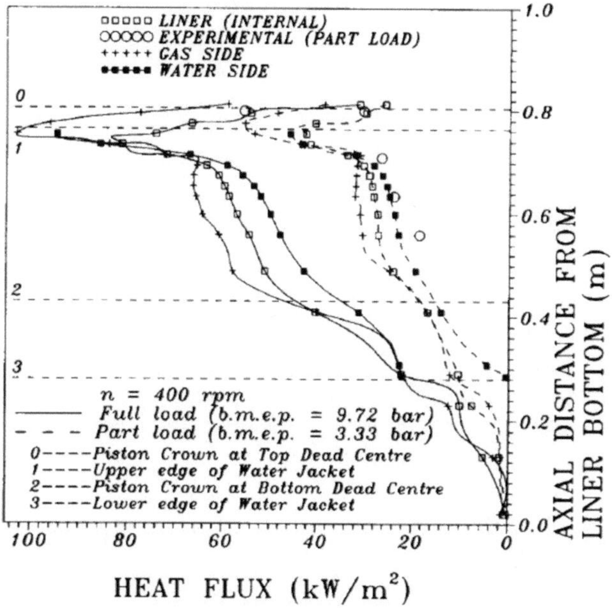

Figure 2.8: Heat fluxes in the cylinder liner of a diesel engine. Two load cases as well as three different surface segments are shown. The experimental line corresponds to the internal segment. Figure reprinted from Rakopoulos and Mavropoulos (1996) (with permission from John Wiley and Sons **https://doi.org/10.1002/(SICI)1099-114X(199605)20:5%3C437::AID-ER169%3E3.0.CO;2-J**).

2.2.7 Piston specific boundary conditions

Beside the heat transfer from combustion, heat dissipation to the cylinder liner and the lubricant is essential. A common strategy to estimate this heat flux is to use heat transfer coefficients or thermal resistances (Esfahanian et al., 2006). According to the model depth, various resistances have to be added. Examples are piston rings or lubricant layers. Regarding a perfect cylindrical geometry, the thermal resistance for an inner surface with radius r_i is given by:

$$R_i = \frac{\ln(r_{i+1}/r_i)}{2\pi H_i \lambda_i}, \tag{2.33}$$

with a corresponding thickness H_i and the thermal conductivity λ_i of the layer. Be aware of the conversion of thermal resistances R_i and heat transfer coefficients $\alpha = 1/\left(R_i A_{\text{eff}}\right)$ with the effective area A_{eff} of the corresponding surface. Exemplary reference values can be found in Mizuno et al. (2009) and Esfahanian et al. (2006). Challenges in modeling oil-jet cooling for automotive pistons are the wide range of oil Prandtl numbers $Pr \approx 100 - 1000$ and various nozzle to piston distances of $\tilde{z}/d_0 \approx 100$, with d_0 as the jet nozzle diameter. Area averaged Nusselt number correlations for one oil jet can be found in Easter et al. (2014). The results are listed in Table 2.3.

Correlation	Conditions
$Nu = 0.066\,(\tilde{z}/d_0)^{-0.034}\,Re_0^{0.541}\,Pr_0^{0.4}$	$90 < Pr_0 < 115$, $1000 < Re_0 < 4500$ (Nondivergent, laminar)
$Nu = 0.028\,(\tilde{z}/d_0)^{-0.034}\,Re_0^{0.652}\,Pr_0^{0.4}$	$90 < Pr_0 < 750$, $100 < Re_0 < 4500$ (Divergence, breakup)

Table 2.3: Area averaged Nusselt number correlations for oil jet piston cooling.

The subscript 0 indicates oil nozzle conditions. Vertical distance should be in the range of $45 < \tilde{z}/d_0 < 95$. In Varghese and Agarwal (2004), a radial distribution of heat transfer coefficients can be found. However, these correlations are only valid for high Reynolds numbers $4500 < Re_0 < 52000$ and low Prandtl numbers $Pr_0 \approx 10$, and are, therefore, questionable for engines. Compared with Table 2.3, an alternative approach is suggested in Esfahanian et al. (2006). In this case, the HTC is modeled according to

$$\alpha = 900\left(n_{\text{engine}}/(4600\,\text{rpm})\right)^{0.35}\ \text{W/m}^2\text{K}, \qquad (2.34)$$

$$\alpha = 240\left(n_{\text{engine}}/(4600\,\text{rpm})\right)^{0.35}\ \text{W/m}^2\text{K}. \qquad (2.35)$$

The first empirical Eq. (2.34) is used for splash cooling types on the crown underside, whereas the second Eq. (2.35) is suggested for skirt undersides. As one can see, Eq. (2.34) and (2.35) does not account for the nozzle conditions, whereas Table 2.3 only takes the oil jet into account.

2.3 Turbulent heat transfer modeling within CFD-CHT method

In the following chapter, a possible set of equations for turbulence modeling with increased heat transfer within RANS simulations is given. The equations are based on incompressible flows, as it is the case in engine water jackets. Boiling effects are not discussed.

2.3.1 Turbulence modeling

With the velocity vector $\boldsymbol{u}$, the inner energy e, and the deviatoric stress tensor $\boldsymbol{\tau}$, the basic conservation equations for mass, linear momentum and energy for incompressible fluids read (Oertel jr. et al., 2011)

$$\operatorname{div}(\boldsymbol{u}) = 0, \qquad (2.36)$$

$$\rho\left(\frac{\partial \boldsymbol{u}}{\partial t} + \operatorname{grad}(\boldsymbol{u})\,\boldsymbol{u}\right) = -\operatorname{grad}(p) + \operatorname{div}(\boldsymbol{\tau}), \qquad (2.37)$$

$$\rho\left(\frac{\partial e}{\partial t} + \boldsymbol{u}\cdot(\operatorname{grad}(e))\right) = -\operatorname{div}(\boldsymbol{q}) + \boldsymbol{\tau}\cdot\cdot\operatorname{grad}(\boldsymbol{u}). \qquad (2.38)$$

In forced convection, the driving force is external to the fluid (Lienhard and Lienhard, 2003), and, therefore, volume forces, such as gravity, are neglected. Additional energy sources, which could model energy releases during chemical reactions, are irrelevant in engine water jackets. For the deviatoric stress tensor $\boldsymbol{\tau}$, the classical approach according to Stokes is given by

$$\boldsymbol{\tau} = 2\mu \operatorname{sym}(\operatorname{grad}(\boldsymbol{u})). \tag{2.39}$$

The heat flux $\boldsymbol{q}$ can be modeled with the Fourier law, which assumes a linear relationship between heat flux and temperature gradient. Using the decomposition of an arbitrary quantity $\psi = \langle \psi \rangle + \psi'$ in its expectation value and its fluctuation regarding time, one gets the Reynolds averaged Navier Stokes equations for incompressible fluid flow by averaging Eq. (2.37). Compared to the initial equation, due to its non-linearity, the extra term $-\rho \operatorname{div}(\boldsymbol{R})$ has to be added. This additional flux of momentum due to turbulence is described by the Reynolds stress tensor

$$\boldsymbol{R} = \langle \boldsymbol{u}' \otimes \boldsymbol{u}' \rangle. \tag{2.40}$$

According to a closed-form expression, the Reynolds stress tensor $\boldsymbol{R}$ can be modeled with the approach by Boussinesq

$$\boldsymbol{R} = \frac{2}{3} k \boldsymbol{I} - 2\nu_{\mathrm{t}} \operatorname{sym}(\operatorname{grad}(\langle \boldsymbol{u} \rangle)), \tag{2.41}$$

with the turbulent viscosity $\nu_{\mathrm{t}}(\boldsymbol{x}, t)$, which has to be modeled. The turbulent kinetic energy is described by $k = 1/2 \langle \boldsymbol{u} \cdot \boldsymbol{u} \rangle$. According to Pope (2010), such a local model can deliver good results if turbulent quantities like gradients of the mean velocity field are not too large. Interestingly, such conditions are often satisfied if the turbulent production and dissipation are in the same range. In industrial CFD applications, due to the fact that gradients cannot be very well resolved within the turbulent boundary layer, the wall laws for the viscous and logarithmic layer

$$u^+ = y^+, \qquad u^+ = \frac{1}{\kappa_k} \ln\left(E_\kappa y^+\right) \tag{2.42}$$

are usually used (Pope, 2010). They are based on equilibrium considerations. The Von-Karman constant is described by E_κ, and the logarithmic wall constant by κ_k. The dimensionless wall distance y^+ and velocity u^+ are defined with the friction velocity u^* according to:

$$y^+ = \frac{u^* y}{\nu}, \qquad u^* = \sqrt{\frac{\tau_{\text{wall}}}{\rho}}, \qquad u^+ = \frac{\langle u \rangle}{u^*}. \tag{2.43}$$

The shear stress at the wall is described by τ_{wall}. Within CFD applications, u^* can be solved iteratively, or is derived from turbulence quantities specific to the particular turbulence model. A possible description of Eq. (2.42) for a continuous and continuously differentiable formulation can be found in Reichardt (1951). Further methods to model near-wall turbulence is the two-layer approach by Rodi (1991), where different algebraic equations are used for the turbulent viscosity ν_t and dissipation ϵ in the cell next to the wall: a possible description for forced convection can be found in Wolfstein (1969). Here, ϵ describes the turbulent dissipation according to $\epsilon = 2\nu\langle \boldsymbol{s} \cdot\cdot \boldsymbol{s}\rangle$, with $\boldsymbol{s} = \text{sym}(\text{grad}\,(\boldsymbol{u}'))$. To model the turbulent viscosity $\nu_t(\boldsymbol{x}, t)$ in Eq. (2.41), the SST k-ω turbulence model by Menter (Menter, 1994) can be used. Most of the equations are described in Siemens (2016). According to Hölz (2015), compared with other turbulence models like the k-ϵ or Spalart-Allmaras approach, this turbulence model offers some advantages regarding turbulent heat transfer. Therefore, balance equations are solved for the two additional field variables k and $\omega = \epsilon/k$. The transport equations for a fixed control volume are given by

$$\frac{\mathrm{d}}{\mathrm{d}t}\int_V \rho k \ \mathrm{d}V + \int_A \rho k \langle \boldsymbol{u} \rangle \ \mathrm{d}\boldsymbol{a}$$
$$= \int_A \left(\mu + \frac{\mu_\mathrm{t}}{\sigma_k}\right) \mathrm{grad}\,(k) \ \mathrm{d}\boldsymbol{a} + \int_V G_k - \rho\beta^* \,(\omega k) \ \mathrm{d}V, \tag{2.44}$$
$$\frac{\mathrm{d}}{\mathrm{d}t}\int_V \rho \omega \ \mathrm{d}V + \int_A \rho \omega \langle \boldsymbol{u} \rangle \ \mathrm{d}\boldsymbol{a}$$
$$= \int_A \left(\mu + \frac{\mu_\mathrm{t}}{\sigma_\omega}\right) \mathrm{grad}\,(\omega) \ \mathrm{d}\boldsymbol{a} + \int_V \left[G_\omega - \beta\rho\omega^2 + D_\omega\right] \ \mathrm{d}V. \tag{2.45}$$

The first equation is a modification of the exact balance equation for the turbulent kinetic energy. The turbulent Prandtl numbers are described by σ_k and σ_ω. G_k is the production term according to

$$G_k = \mu_\mathrm{t} S^2 - \frac{2}{3}\rho k \mathrm{div}\,(\langle \boldsymbol{u} \rangle) - \frac{2}{3}\mu_\mathrm{t}\,(\mathrm{div}\,(\langle \boldsymbol{u} \rangle))^2 . \tag{2.46}$$

In case of the turbulent frequency ω, two additional production terms are given by G_ω and D_ω. Because of mass conservation, the modulus of the mean strain rate tensor $S = \sqrt{2\boldsymbol{S} \cdot\cdot \boldsymbol{S}}$, with $\boldsymbol{S} = \mathrm{sym}(\mathrm{grad}\,(\langle \boldsymbol{u} \rangle))$, is the only non-zero term. According to Siemens (2016), the empirical parameter β^* is set to 0.09. β is a further model parameter which is specified in Table 2.4. The second Eq. (2.45) is a modification of the standard k-ω turbulence model by Wilcox. In his approach, he neglects the cross diffusion term "$\propto \mathrm{grad}\,(\omega) \cdot \mathrm{grad}\,(k)$", which arises from derivations of the k-ϵ model proposed by Launder (1972). This idea seems to be advantageous for regions which are close to the boundary with flow separation and pressure gradients. However, because of some observed deficits in flows more far away from the boundary, the SST k-ω turbulence model was proposed. The idea is to combine both advantages as a function of the wall distance. A more detailed overview about the derivations can be found in Hölz (2015). Both production terms are given by

$$G_\omega = \rho\gamma \left[S^2 - \frac{2}{3} \left(\mathrm{div}\left(\langle \boldsymbol{u} \rangle\right)\right)^2 - \frac{2}{3}\omega \mathrm{div}\left(\langle \boldsymbol{u} \rangle\right) \right], \tag{2.47}$$

$$D_\omega = \rho \frac{2}{\omega} \sigma_{\omega 2} \left(1 - F_1\right) \mathrm{grad}\left(k\right) \cdot \mathrm{grad}\left(\omega\right). \tag{2.48}$$

Again, only the first term in the production term G_ω is non-zero. Details to the two model parameters γ and $\sigma_{\omega 2}$ can be found in Table 2.4. F_1 is a smoothing function, which is one at the wall and goes to zero with increasing wall distance. By this means, the already mentioned cross-diffusion term in Eq. (2.48) gains in importance far away from the boundary. Details to the exact implemented function F_1 can be found in the user manual by Siemens (2016). According to a dimensional analysis, the modified turbulent viscosity is modeled according to

$$\nu_\mathrm{t} = k\tilde{T}. \tag{2.49}$$

For the turbulent time scale $\tilde{T}$, following expression is used:

$$\tilde{T} = \frac{1}{\max\left(\omega, SF_2/a_1\right)}. \tag{2.50}$$

According to Siemens (2016), the model parameter a_1 is given by 0.31. Again, the function F_2 is a smoothing function, which is one at the wall and goes to zero with increasing wall distance. This modified approach within the SST model ("**S**hear **s**tress **t**ransport") is expected to improve flow simulations with reversed pressure gradients and subsequent separation phenomena. In the original paper by Menter (1994), the amount of the rotation tensor is selected instead of $\boldsymbol{S}$. For the above used model parameters, following approach is used:

$$\phi = F_1\phi_1 + \left(1 - F_1\right)\phi_2. \tag{2.51}$$

Model parameter ϕ	**Constant** ϕ_1	**Constant** ϕ_2
β	$\beta_1 = 0.075$	$\beta_2 = 0.0828$
γ	$\gamma_1 = \frac{\beta_1}{\beta^*} - \sigma_{\omega_1} \frac{\kappa_k^2}{\sqrt{\beta^*}}$	$\gamma_2 = \frac{\beta_2}{\beta^*} - \sigma_{\omega_2} \frac{\kappa_k^2}{\sqrt{\beta^*}}$
σ_k	$\sigma_{k1} = 0.85$	$\sigma_{k2} = 1.0$
σ_ω	$\sigma_{\omega_1} = 0.5$	$\sigma_{\omega 2} = 0.856$

Table 2.4: Calibration of the SST k - ω model.

According to Siemens (2016), possible values are listed in Table 2.4. The variation is the result of the previously mentioned approach to combine the original k-ϵ and k-ω model by Wilcox. For the wall shear stress velocity $u^* = \sqrt{\tau_{\text{wall}}/\rho}$, following approach is used:

$$u^* = \sqrt{\frac{\tilde{g}\nu\langle u\rangle}{y} + (1-\tilde{g})\sqrt{\beta^*}k}. \tag{2.52}$$

Again, $\tilde{g}$ is an empirical smoothing function of the form

$$\tilde{g} = \exp\left(-\frac{Re_y}{11}\right), \tag{2.53}$$

which depends on the wall distance y according to the local Reynolds number $Re_y = \sqrt{k}y/\nu$. Remember that the wall shear stress velocity is used for the non-dimensional form of u^+. Therefore, an alternative approach would be the iterative determination of u^*. Due to the difficulty in resolving the boundary layer, an extra expression for the production term G_k in the first cell next to the wall can be used:

$$G_k = \tilde{g}\mu_t S^2 + (1-\tilde{g})\frac{1}{\mu}\left(\rho u^* \frac{u}{u^+}\right)^2 \frac{\partial u^+}{\partial y^+}. \tag{2.54}$$

The two boundary conditions for the additional field functions k and ω are results of diverse experimental or theoretical investigations. For the turbulent kinetic energy, the derivative at the wall can be set according to

$$\left.\frac{\partial k}{\partial n}\right|_{w} = 0. \tag{2.55}$$

Equation (2.55) is motivated by direct numerical simulations performed for a channel flow with $Re = 13.750$ (Pope, 2010). For the turbulent frequency ω, following expression is used for the cell near to the wall:

$$\omega = \tilde{g}\frac{6\nu}{\beta y^2} + (1-\tilde{g})\frac{u^*}{\sqrt{\beta^*}\kappa_{\mathrm{lo}} y}. \tag{2.56}$$

More investigations about boundary conditions for diverse turbulent quantities can be found in Hadzic (1999) or Ferziger and Peric (2002).

2.3.2 Increased heat transfer due to turbulence

Increased heat flux due to turbulence can be mathematically described with the non-linear term in the energy Eq. (2.38). Normally, dissipation due to velocity gradients can be neglected. In relation to the constitutive relation by Fourier, the additional correlation term is often modeled according to

$$-\operatorname{div}\left(\langle \boldsymbol{u}'T'\rangle\right) = \operatorname{div}\left(\rho c_p a_{\mathrm{t}} \operatorname{grad}\left(\langle T\rangle\right)\right). \tag{2.57}$$

For the turbulent temperature conductivity $a_{\mathrm{t}}(\boldsymbol{x}, t)$, the turbulent Prandtl number is defined by

$$Pr_{\mathrm{t}} = \frac{\nu_{\mathrm{t}}}{a_{\mathrm{t}}}. \tag{2.58}$$

In accordance with the physical Prandtl number, a constant value of 0.9 for water channels are used. A very good overview about different approaches for turbulent Prandtl numbers can be found in Hasan (2007). Another, more computationally intensive, approach for modeling turbulent heat fluxes can be found in Kenjeres et al. (2005). Two additional transport equations are solved there. Using the first equation of

(2.42) for the viscous sublayer, one can derive following dimensionless temperature profile

$$T^+ = \frac{(T - T_{\text{wall}})\, \rho c_p u^*}{q_{\text{wall}}} = Pr\, u^+ = Pr\, y^+. \tag{2.59}$$

For its derivation, the simplified two-dimensional, stationary energy equation with $a = \lambda/\rho c_p$ is used:

$$\langle u_1 \rangle \frac{\partial \langle T \rangle}{\partial x} + \langle u_2 \rangle \frac{\partial \langle T \rangle}{\partial y} = \frac{\partial}{\partial y} \left(a \frac{\partial \langle T \rangle}{\partial y} - \langle T' v' \rangle \right). \tag{2.60}$$

In addition, one has to neglect the increase of heat flux due to turbulence from Eq. (2.57). The corresponding wall quantities are T_{wall} and q_{wall}. The same procedure can be done for the logarithmic region using the second equation of (2.42), and assuming that the total heat flux $\boldsymbol{q}$ is dominated by the turbulent contribution $\langle \boldsymbol{u}' T' \rangle$. Therefore, the turbulent viscosity can be modeled with the Prandtl mixture path model: details can be found in Merker and Baumgarten (2000). The resulting dimensionless temperature profile can be written as follows:

$$T^+ = \frac{(T - T_{\text{wall}})\, \rho c_p u^*}{q_{\text{wall}}} = Pr_{\text{t}} \left(u^+_{\text{turbulent}} \right) + \text{const}(Pr). \tag{2.61}$$

For its derivation, as a common approach for the turbulent length scale l in the logarithmic layer, a linear model $l = \kappa_k y$ is used. The velocity profile in the logarithmic region is $u^+_{\text{turbulent}}$. The constant const(Pr) results from the integration procedure, and has to be a function of the Prandtl number. This ensures the thermal resistance of the viscous sublayer. Following approach can be used:

$$\text{const}(Pr) = Pr_{\text{t}} 9.24 \left[\frac{Pr}{Pr_{\text{t}}}^{3/4} - 1 \right] \left[1 + 0.28 \exp\left(\frac{-0.007 Pr}{Pr_{\text{t}}} \right) \right]. \tag{2.62}$$

Chapter 3

Influence of various engine settings on stationary, cycle averaged component temperatures[1]

3.1 Outline of the chapter

The research question can be formulated as follows: with regard to averaged solid temperatures, knowing full well that instantaneous in-cylinder heat transfer is a highly complex problem, is it possible to use simplified, statistical calculation methods without simulating crank angle resolved heat fluxes? In particular, with respect to these integral heat fluxes, how purposeful are fast and simplified heat transfer models regarding to different engine settings? Previous investigations of the works mentioned in the introduction show that the instantaneous heat flux of these models can strongly differ from each other and from experimental measurements. Therefore, this chapter complements existing published works which studied crank angle resolved heat fluxes or temperature swings rather than averaged temperatures.

[1] This chapter is based on the paper "CFD-CHT calculation method using Buckingham Pi-Theorem for complex fluid-solid heat transfer problems with scattering boundary conditions" (Hölz et al., 2018a) **DOI: https://link.springer.com/article/10.1007/s41104-018-0026-z**

In relation with the statistical description and calculation method, cyclical fluctuations in the pressure curves are measured and described statistically using probability density functions: gas pressure, corresponding heat transfer coefficients, cylinder gas temperature, and their product, in particular, are described as random variables. This interpretation enables a time scale separation with a low-pass filter function.
Different operating conditions like varying air-fuel ratios, adjustments of ignition times, or variations in boost pressure are, therefore, experimentally investigated with a 0.5 liter turbocharged single cylinder SI engine and compared to simulation results. In accordance with Chapter 2, the pressure signal from the combustion chamber is used to approximate the heat transfer to different components like cylinder head, piston, valves, or the cylinder liner. With the help of the simplified methods using the Buckingham Pi-Theorem, like correlations after Woschni (Woschni, 1967), Hohenberg (Hohenberg, 1979), or Eichelberg (Eichelberg, 1939), the averaged heat transfer coefficient can then be approximated. The spatial distribution is based on the results presented in Chapter 2.

3.2 Method used in this chapter

The aim of the present chapter is the determination of engine temperatures in dependence of the operating conditions, in particular of the engine application data version. That means different engine mappings like ignition time or air-fuel ratio variations are investigated with regard to solid temperatures. One distinguishes, therefore, between *inner* and *outer* boundary conditions for an engine. The outer boundary conditions can be defined as a time-dependent five-dimensional engine state matrix:

$$\underline{M}(t) = \left(n_{\text{engine}}(t), m_{\text{air}}(t), t_{\text{int}}(t), T_{\text{i}}(t), m_{\text{fuel}}(t)\right). \tag{3.1}$$

In this case, however, the special case of stationary engine states is considered. The entries are engine speed, air mass flow rate, inlet air temperature, induced torque, and fuel mass flow rate. Obviously, these

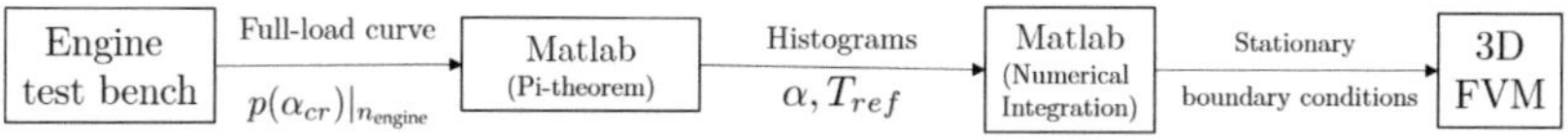

Figure 3.1: Method overview for generating boundary conditions for a detailed three-dimensional CFD-CHT simulation with the help of pressure indication measurement $p(\alpha_{cr})$ at a given engine speed n_{engine}.

outer boundary conditions have to be seen as *concentrated parameters* that means they are no functions of place. Assuming that this state matrix is known, one has to determine the inner boundary conditions. In the context of conjugate heat transfer simulations, this includes the determination of thermal boundary conditions: temperature (Dirichlet BC), heat flux (Neumann BC), or heat transfer coefficients (Robin BC). In the following, a detailed three-dimensional CFD-CHT simulation is presented. A method is suggested how one can translate *outer, concentrated* boundary conditions into *inner, distributed* boundary conditions of an engine. *Distributed* means that these BCs have to be functions of place. To control the outer BCs to a certain point, lots of variables are available: injection and ignition time or boost pressure, only to name a few. The inner BCs, for example, are results of the resulting turbulent flow of the water cooling system or the turbulent flow with chemical reactions in the cylinder. Especially, this internal flow is subjected to scattering heat releases and pressure developments.
An overview of the method is shown in Fig. 3.1. An own implemented code in MATLAB© is used to determine boundary conditions for the subsequent detailed 3D-FVM simulation. One needs, therefore, a high-pressure indication measurement $p(\alpha_{\text{cr}})$ of the engine for the used application data version. It is important to notice that lots of information about heat transfer is hidden in the pressure signal which is resolved

with 0.1 °CA. Most of the following heat transfer boundary conditions are Robin BC. Assuming that all internal flows can be described by forced convection, one can assume a heat flux of the Newton form with temperature independent heat transfer coefficient α:

$$\boldsymbol{q} = -\alpha\,(T_{\text{ref}} - T_s)\,\boldsymbol{n}. \tag{3.2}$$

The idea is to interpret these quantities as random variables. In this case, however, one has to use conditional probability density functions. In the case of the heat transfer coefficient, this is $p_{\alpha|\underline{M}}$ for a given engine state $\underline{M}$. Because one can take into account the strong scattering nature of the cylinder pressure, the description as a random variable is quite useful. Especially, combustion with high λ_{cmb} values tends to scatter. λ_{cmb} describes the ratio between actual air mass and stoichiometric air mass. For calculating stationary temperature fields, one has to integrate the approximated density functions to get expectation values of heat transfer coefficients and the corresponding reference temperatures. Strictly spoken, as a result of the oscillating four stroke process, an engine does not have a stationary state. Nevertheless, one can consider quasi-stationary conditions if one imagines a time scale separation realized by a statistical description. This is a great advantage because of the non-linearity in Eq. (3.2):

$$\begin{aligned}
\langle \boldsymbol{q} \rangle &= -\langle \alpha\,(T_{\text{ref}} - T_s)\rangle \boldsymbol{n} \\
&= -\left(\langle \alpha T_{\text{ref}}\rangle - \langle \alpha T_s\rangle\right)\boldsymbol{n} \\
&= -\langle \alpha \rangle \left(\frac{\langle \alpha T_{\text{ref}}\rangle}{\langle \alpha \rangle} - \langle T_s \rangle\right)\boldsymbol{n} \\
&= -\langle \alpha \rangle \left(T_{\text{mod}} - \langle T_s \rangle\right)\boldsymbol{n}.
\end{aligned} \tag{3.3}$$

Note that $\langle\cdot\rangle$ is a mathematical operator for the expectation value. $\langle T_s \rangle$ is the searched solid temperature which can be a function of time in transient simulations. The term $T_{\text{mod}} = \langle \alpha T_{\text{ref}}\rangle / \langle \alpha \rangle$ will be called the modified reference temperature. Whenever it will be referred to a reference temperature, this modified version is meant. Notice that Eq. (3.3) can be rewritten as

$$\begin{aligned}\langle \boldsymbol{q} \rangle &= -\langle \alpha \left(T_{\text{ref}} - T_s \right) \rangle \boldsymbol{n} \\ &= -\left(\langle \alpha T_{\text{ref}} \rangle - \langle \alpha T_s \rangle \right) \boldsymbol{n} \\ &= -\left(\langle \alpha \rangle \langle T_{\text{ref}} \rangle + \langle \alpha' T'_{\text{ref}} \rangle - \langle \alpha T_s \rangle \right) \boldsymbol{n}.\end{aligned} \tag{3.4}$$

Therefore, the modified gas temperature T_{mod} implies the effect due to the correlation $\langle \alpha' T'_{\text{ref}} \rangle$. The superscript $(\cdot)'$ describes fluctuations. In Eq. (3.3), it is assumed that the correlation of the heat transfer coefficient α and the solid temperature T_s is negligibly small:

$$\langle \alpha' T'_s \rangle \ll \langle \alpha \rangle \langle T_s \rangle. \tag{3.5}$$

Because of large heat capacities of solids, their temperatures behave much slower than variations of heat transfer coefficients. Inequality (3.5) can be checked with the following conservative assumptions: temperature fluctuations T'_s and HTC fluctuations α' are inphase with a maximum temperature amplitude of 30 K during an engine cycle (Kenningley and Morgenstern, 2012). For typical engine operating conditions with solid temperatures in the range of 400-500 K, the left-hand side of Eq. (3.5) is two orders of magnitude smaller than the right-hand side. In the subsequent FVM simulation, $\langle \alpha \rangle$ and T_{mod} are input parameters describing Robin boundary conditions. For the correlation of the HTC and ACT in the combustion chamber, it follows:

$$\langle \alpha' T'_{\text{ref}} \rangle = \langle \alpha T_{\text{ref}} \rangle - \langle \alpha \rangle \langle T_{\text{ref}} \rangle. \tag{3.6}$$

Of course, this correlation is not zero and, hence, cannot be neglected in the case of the combustion chamber. Pressure and temperature peaks occur simultaneously. For this reason, the HTC and the ACT have a strong correlation. Summarizing the above-mentioned statistical description with the example of the HTC in the combustion chamber, one gets for any arbitrary random variable $\tilde{f}(\alpha)$ the following expression for its expectation value:

$$\langle \tilde{f}(\alpha) \rangle = \int\limits_{\mathbb{R}_{\geq 0}} \tilde{f}(A) p_{\alpha|\underline{M}}(A) \, dA. \tag{3.7}$$

In this case, the variable A is the realization of the random variables α describing the HTC for a given engine state matrix $\underline{M}$. $p_{\alpha|\underline{M}}(A)$ is the conditional PDF. The probability density functions are implemented as normed histograms. For the numerical integration, trapezoidal rule is chosen. Concerning the simulation method, a combination of a detailed CFD RANS simulation for the water channel and a simplified approach using the aforementioned Buckingham Pi-Theorem is used. The last one offers boundary conditions for the surfaces which are wetted with gas. According to a CHT method, the solid and fluid regions within the FVM simulation are solved simultaneously.

3.3 Determination of boundary condition

3.3.1 Heat transfer modeling in the inlet and outlet systems

The correlation according to Eq. (2.11) is used for the outlet valve stem. The constant is $C_1 = 1.84$. For the inlet valve stem, the coefficient in Eq. (2.11) has to be reduced by 40 %. For the intake port, the simple form (2.15) is chosen. According to a fully developed flow, in this chapter, a value of one is used for the exponent m. The coefficient c serves as a calibration parameter. For the exhaust port, the model according to Eq. (2.12) is proposed. Following the original paper, the constant C_2 is set to $C_2 = 8/\pi$. Gas temperature measurements in the intake and exhaust port serve as input parameters.

3.3.2 Heat transfer modeling in the combustion chamber

Regarding the sensitivity to different engine mappings, three different models are investigated. Radiation is generally neglected. Depending on the channel geometry swirl, flow structures can occur in engine cylinders. It is important to note that the investigated engine does not have any significant swirl motion. In Table 2.2, a summary of the investigated models is given. According to Fig. 3.6, the model parameters $C_{\text{Eichelberg}}$, C_{Woschni} and $C_{\text{Hohenberg}}$ are calibration values.

3.4 Finite volume model

The commercial software StarCCM+ by Siemens (release v11.06) is used for calculating solid temperatures according to a CHT simulation. The model contains diverse engine components, e.g., the crankcase, cylinder head, cylinder liner, piston, cam cover, valves as well as the valve rings.

3.4.1 Mesh study

Regarding several thermal quantities, mesh studies are performed to ensure mesh independence. First, the local mesh size in the cylinder head is varied in the region of the high speed channel flow around the valve seats and cylinder liner cooling areas. Second, the local mesh size of the water region is varied at the same domain. The boundary layer, however, is kept constant within three cells. The first cell lies either in the viscous sublayer (dimensionless wall distance $y^+ < 5$), or in the logarithmic layer (dimensionless wall distance $y^+ < 35$). The results are shown in Figs. 3.3. An overview about the mesh can be seen in Fig. 3.2.

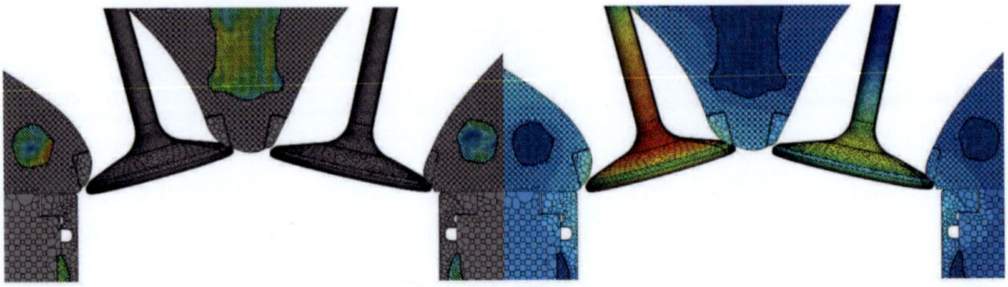

Figure 3.2: Overview about local mesh size. Left: qualitative fluid velocity field in the water channel. Right: qualitative temperature field for fluid and solid regions.

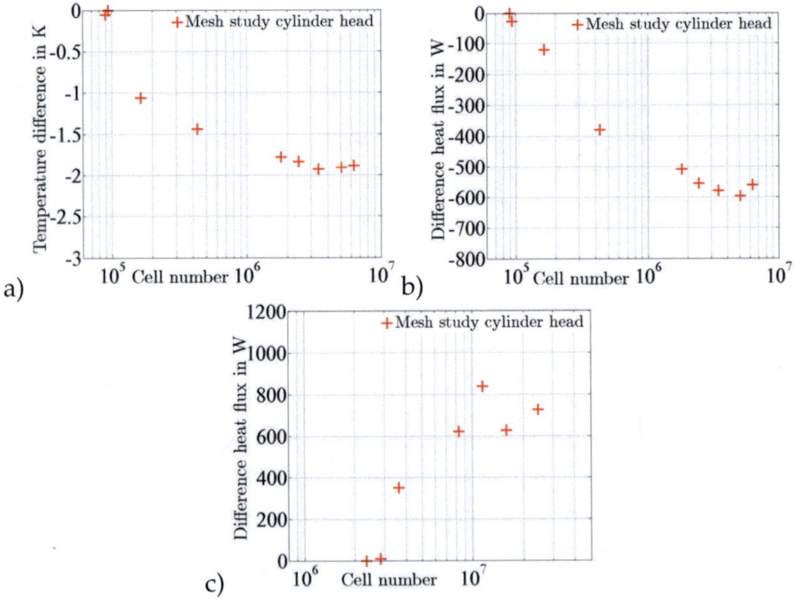

Figure 3.3: Mesh study for the cylinder head. Local mesh is varied in the region of the high speed channel flow around the valve seats and cylinder liner cooling areas. a) Solid region: surface average solid temperature at the combustion chamber wall. b) Solid region: heat flux of the wetted surfaces. c) Water region: heat flux of the wetted surfaces.

3.4.2 Spatial distribution of heat transfer coefficients and turbulence modeling

To take account of the spatial distribution of the HTC within the combustion chamber, detailed 3D-CFD simulation results from Mohammadi et al. (2008) shown in Fig. 2.3 used. The relative distribution between different components is listed in Table 3.1. In addition, the spatial dependence of the HTC on the cylinder head and piston surface is assumed as a linear function with respect to the radial position. Regarding the liner, a linear distribution with respect to the axial position is modeled. Therefore, only the part is considered which is available for the propagating flame. The radial and axial distributions in the combustion chamber are implemented in such a way that the spatial course is continuous. It is important to note that the global HTC, according to Eq. (5.3), serves as a preexponential factor for each component.

Component	**Weighting factor**
Cylinder liner	1.0
Cylinder head	3.25
Piston	2.5
Intake and exhaust valves	3.6

Table 3.1: Relative distribution of the HTC between different components in the combustion chamber (Mohammadi et al., 2008).

Special attention is required for the contact pair valve ring (inlet and exhaust side). For the thermal resistance, respectively, heat transfer coefficient, a value of 35000 W/(m^{2}K) is used (Fieberg and Korthäuer, 2006). An analogous adaption of the thermal resistance is necessary for the contact pair valve-valve guide. The empirical value of 170 W/(m^{2}K) is, therefore, chosen (Shojaefard et al., 2005). Increased heat transfer due to turbulence is calculated according to Section 2.3.

3.4.3 Spark plug modeling

The spark plug is modeled as a perfectly isolated steel layer on the one side which has a perfect heat conduction to the cylinder head on the other side. At the bottom surface which projects into the combustion chamber, the heat transfer is also given by a heat transfer coefficient and the average gas temperature. However, due to the fact that the spark plug projects deeper into the combustion chamber, one has to modify the heat transfer coefficient resulting from the Pi-Theorem. A heat increasing factor α_{increase} is, therefore, introduced resulting in a modified heat transfer coefficient for the spark plug α_{spark}:

$$\alpha_{\text{spark}} = \alpha_{\text{increase}} \alpha_{\pi}. \tag{3.8}$$

α_{π} is the heat transfer coefficient resulting from the Pi-Theorem. The factor α_{increase} is calibrated once.

3.5 Results and discussion

A 0.5 liter turbocharged single cylinder SI engine is investigated under varying boundary conditions. Fig. 3.4 gives an overview about the parameter set and the corresponding in-cylinder pressure curves. Type K Thermocouples with an uncertainty of 1 K are used for the measurement of solid temperatures. Heat-conducting paste with a thermal conductivity of 2.5 W/(mK) is put between the solid surface and the thermocouples. Fluid temperatures are measured with Pt100 sensors which have an uncertainty of 0.053 K. Measuring points are installed inside of the spark plug coat, at the combustion chamber wall of the cylinder head and the exhaust valve seat (seat ring). A thermocouple has, therefore, been welded into the spark plug which ends at the last thread. For the combustion chamber wall, a hole until 1 mm above the surface is drilled.

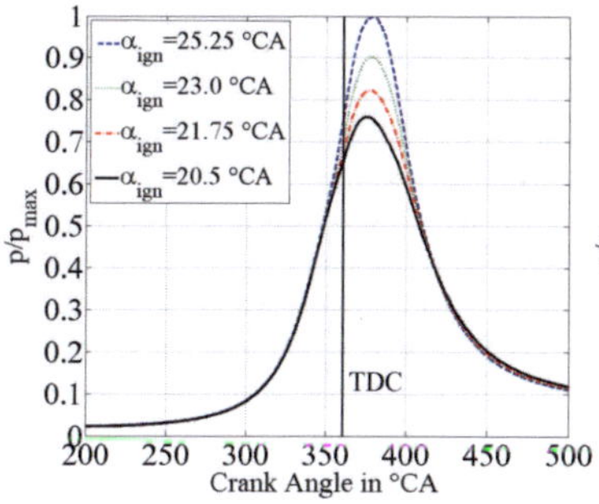

(a)

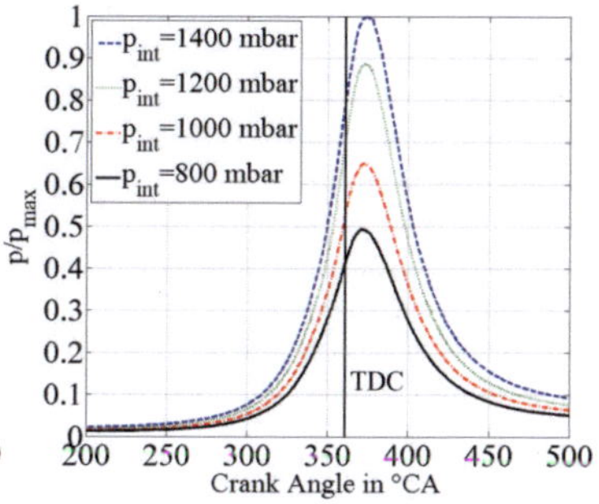

(b)

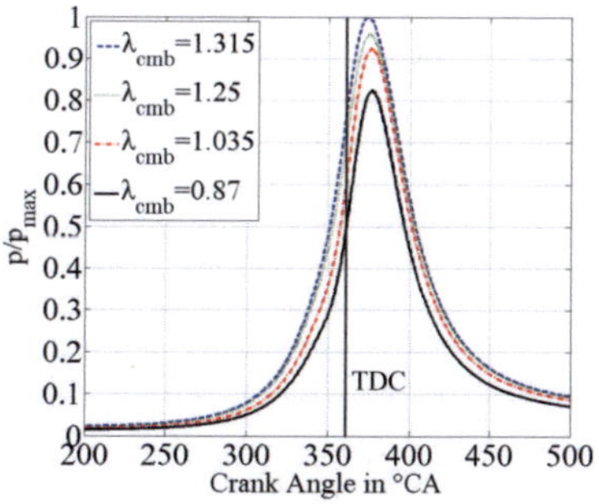

(c)

Figure 3.4: Averaged in-cylinder pressure curves for three different thermodynamic variations. Within one variation, the pressure curves are normed to the maximum value. All three variations represent full load conditions. (a) Ignition time variation with constant fuel mass flow and air-fuel ratio: $n_{\text{engine}} = 6000$ rpm. (b) Boost pressure variation with constant air-fuel ratio. Ignition time is set for an optimum thermal efficiency: $n_{\text{engine}} = 8000$ rpm. (c) Lambda variation with constant fuel mass flow. Ignition time is set for an optimum thermal efficiency: $n_{\text{engine}} = 8000$ rpm.

The position is located at the outer diameter of the combustion chamber. For each measuring point, 60 combustion cycles are recorded with the help of a piezo-quartz pressure transducer mounted on the cylinder head. A sampling rate of 0.1 °CA is, therefore, used. The crankshaft position is measured by a crankshaft angle sensor to determine the in-cylinder gas pressure as a function of crank angle. Regarding the question whether 60 combustion cycles are sufficient to calculate a representative heat transfer, Fig. 3.5 shows the convergence behavior of the relevant variables. Therefore, following the recognition of Fig. 7.17, the more conservative, strongly scattering spark plug configuration is used. Concerning the following variations of the engine setting, e.g. Fig. 3.8, 3.10, 3.12 and 6.2, 60 cycles seem to be a good compromise between calculation time and a certain safety.

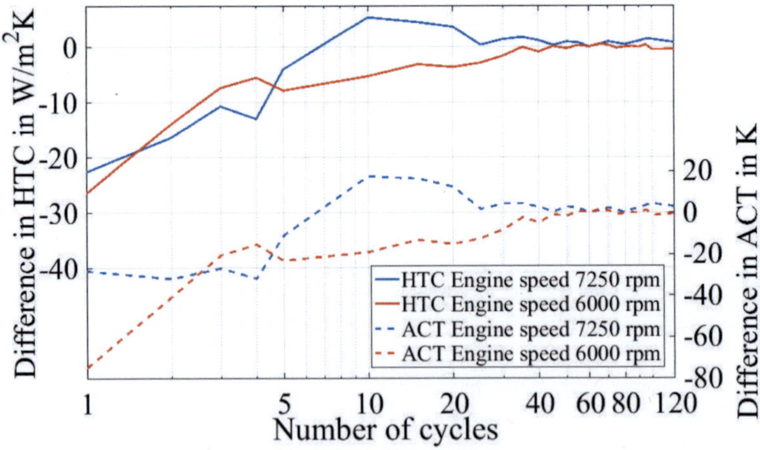

Figure 3.5: Convergence behavior of thermal variables for two different engine speeds. The differences are referred to 60 combustion cycles.

An overview about the method validation is given in Fig. 3.6. The simulated solid temperature difference between different engine settings, which is based on the thermal boundary conditions of the suggested calculation technique and various heat transfer models, is compared with the measured temperature difference.

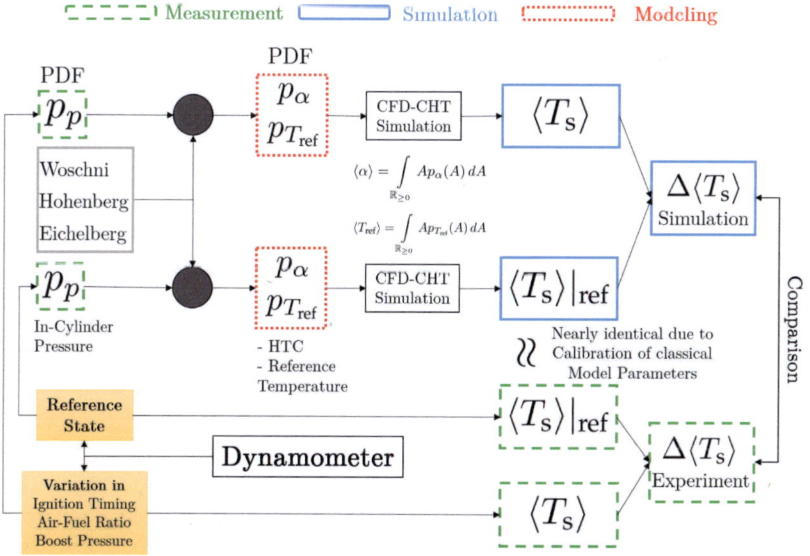

Figure 3.6: Overview about the method validation.

3.5.1 Statistics of maximum in-cylinder pressure

A brief summary of pressure statistics is given in Fig. 3.7. A Gaussian distribution is, therefore, assumed for the maximum pressure p_{max}:

$$p_{p_{max}} = \frac{1}{\sqrt{2\pi\sigma^2}} \exp\left(-\frac{(p_{max} - \xi)^2}{2\sigma^2}\right). \quad (3.9)$$

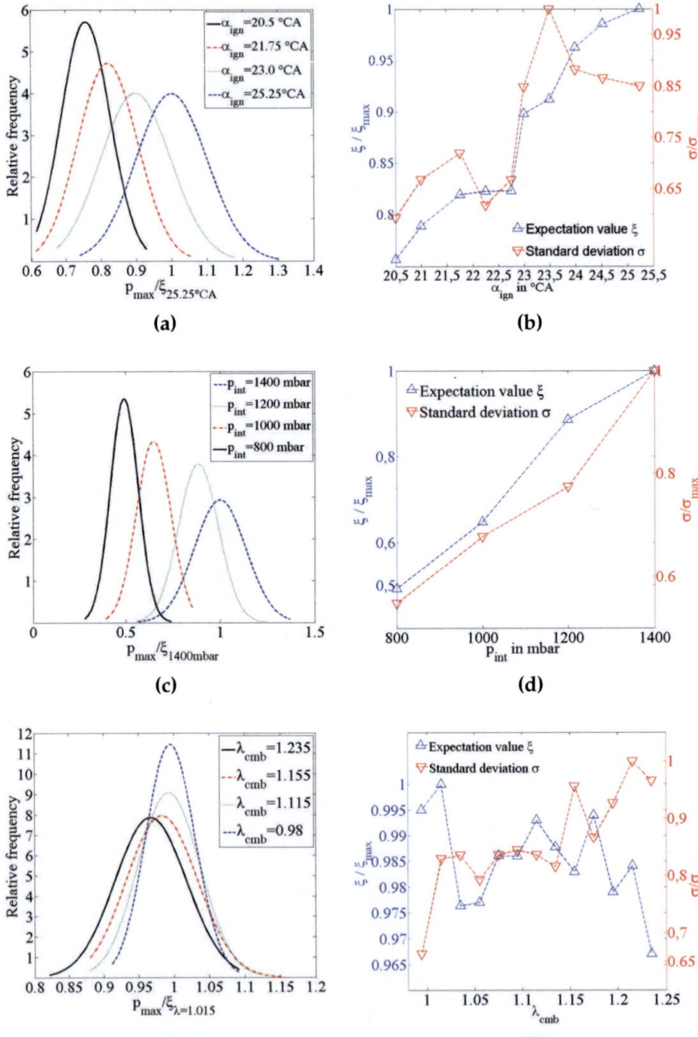

Figure 3.7: Statistics of maximum in-cylinder pressure. Gaussian distribution on the left side. Pressure values are normed to the maximum average value. Expectation value ξ and standard deviation σ on the right side. Within one variation ξ and σ are normed to the maximum value. (a)-(b) Ignition time variation. (c)-(d) Boost pressure variation. (e)-(f) Lambda variation.

According to Table 2.2 and a model constant of $m = 0.8$, the maximum HTC is approximately linear in the maximum pressure. Remember that the average gas temperature $\overline{T}_g$ is nearly proportional to the pressure for a constant amount of molecules which is the reason why the maximum heat flux is approximately in quadratic proportion to the maximum pressure (Eq. (3.2)). Regarding mechanical loads in the high cycle fatigue regime, the maximum pressure is a representative quantity. Details are given in Chapter 7. Note that ignition time and boost pressure variations have a strong effect on the expectation values ξ and standard deviations σ, respectively. For different loads, by adjusting boost pressure, a quite linear behavior can be seen. In contrast to that, lambda variations have only a meaningful effect on σ. The average maximum pressure only reduces about 4 % with increasing lambda values. Because of current racing regulations with a limited maximum fuel flow, it is important to increase the thermal efficiency for optimizing the overall performance. Only conditions with excess air are, therefore, investigated. Details of engine conditions are summarized in the following chapters.

3.5.2 Ignition time variation

There are lots of other reasons to change the ignition angle during a race: in response to different torque requirements because of traction control or due to preceded knocking events, for example. Increasing advance, the thermal efficiency of the engine increases, because more fuel is chemically converted at the top dead center. This results, however, in a higher heat transfer rate to the combustion chamber wall (Fig. 3.4 (a)). In Table 4.1, the most important engine parameters are summarized. t_{amb} and α_{ign} are the ambient air temperature and ignition angle, respectively. The oil and water inlet temperatures are described by $T_{\text{o,i}}$ and $T_{\text{w,i}}$.

Engine parameter	Value	Engine parameter	Value
n_{engine}	6000 rpm	$T_{\text{o,i}}$	388.15 K
t_{int}	312.15 K	$T_{\text{w,i}}$	387.5 K
t_{amb}	294.15 K	m_{fuel}	115.5 mg/stroke
λ_{cmb}	1.2	α_{ign}	20.5 - 25.25 °CA

Table 3.2: Engine parameters during the variation of ignition angle.

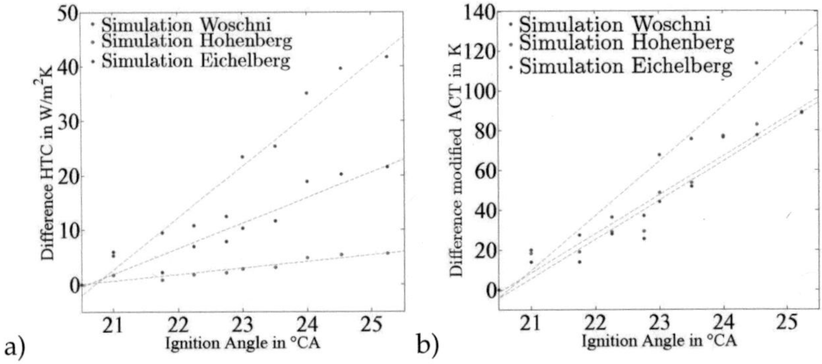

Figure 3.8: Progressions of the heat transfer coefficients and the modified average gas temperatures: a) HTC increase in the combustion chamber due to increased spark advance for three different models. b) Corresponding ACT rise with increasing ignition angle.

For this experiment, two different positions are investigated: the spark plug and the combustion chamber wall.In Fig. 3.8, the variations of the heat transfer coefficients and the modified average gas temperatures are shown. The dashed lines are linear regression lines for a better comparison. Remember that due to the non-linearity of Eq. (3.3), one has to use modified average gas temperatures: $\langle \alpha T_{ref} \rangle / \langle \alpha \rangle$. These values are not independent on the models due to the influence of the heat transfer coefficient α. The most significant differences can be seen in Fig. 3.8 a): whereas the model according to Hohenberg does not change significantly the heat transfer coefficients, there is a huge increase predicted by the Woschni model. According to Fig. 3.8 b), the change in the modified average gas temperature is quite similar for all three models. There is a slightly higher increase for the Woschni model. In Fig. 3.9, the simulation and experimental results are shown for the solid temperatures.

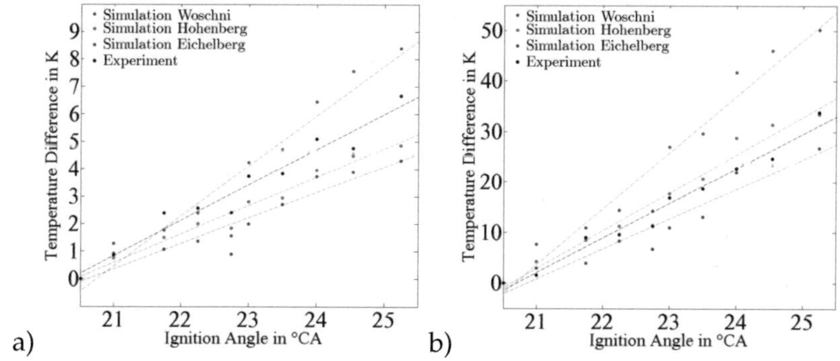

Figure 3.9: Comparison between simulation and experimental results for ignition time variation: a) temperature increase at the combustion chamber wall due to more spark advance. b) Corresponding temperature rise at the spark plug with increasing ignition angle.

The lowest ignition angle point is used for calibration. Compared with variations in the air-fuel ratio or the boost pressure, the temperature differences are small. This is in accordance with Demuynck et al. (2011): varying ignition time caused a time-delayed heat flux, whereas the temperature differences were rather secondary. In addition, a linear regression line is added for each model and the experimental result to evaluate the sensitivities to ignition time variations. Crank angle-based measurements show that the model according to Woschni tends to overpredict the heat transfer at the top dead center region, but underestimates the heat flux during compression and exhaust phases (Sanli et al., 2009). According to Finol and Robinson (2006), Hohenberg's model reveals some improvements in this aspect. Analogous comparisons done by Adair et al. (1972) show that the Eichelberg model predicts lower values than the Woschni model. According to measurements done by Wimmer et al. (2000), one can summarize that the Woschni model predicts higher time averaged heat fluxes and total heat releases for most engine states in comparison with the Hohenberg model. This ratio is a function of engine load and speed. Regarding crank angle-based heat flux measurements, one can guess that the Woschni model overpredicts the time averaged heat flux. Hohenberg's model seems to be better. This tendency, however, does not provide information about heat flux expectation values when using calibrated model equations. The results in Fig. 3.9 show that the Woschni model overpredicts the sensitivity to ignition angle variations for both, combustion chamber wall and spark plug. The other two models give too low gradients for the combustion chamber wall temperatures in comparison with the experimental values. True values probably lie somewhere in between. The temperatures for the spark plug are, however, in good agreement with the measurements. It must be pointed out that the spark plug modeling according to Eq. (3.8) is a strong simplification. It is also possible that this model overpredicts the heat increase to the spark plug in general: a slightly underestimation of the HTC would then result in a good agreement for the spark plug heat transfer. One interesting observation is the noticeable temporary decrease of both temperatures at an ignition angle of 22.75 °CA. It seems that this phenomenon is not a measuring error, because every model predicts this temperature drop. It is satisfying that the models can predict such an unpredictable event using the information hidden in the pressure signal. Remarkably, the power of the engine does not decrease at this measuring point: the positive gradient of thermal efficiency with increasing ignition angle does not show any abnormality. In comparison with Sanli et al. (2009), the predicted differences in the *averaged* HTCs are correctly lower than the corresponding peak values during an engine cycle. For example, consider the prediction by Woschni which gives a difference of the averaged value in the order of 45 W/(m^{2}K). The reported peak value is about 75 W/(m^{2}K). Both results are based on an ignition time interval of 5 °CA. Compared with the Hohenberg model, the resulting difference is only about 6 W/(m^{2}K).

3.5.3 Boost pressure variation with constant lambda

In this experiment, the fuel mass flow and the air mass flow are varied keeping the air-fuel ratio constant. The ignition time is adjusted manually to the individual knocking limit. This experiment can be seen as a load variation at a constant engine speed. The boost pressure is varied between 0.8 and 1.4 bar resulting in a throttled and turbocharged engine state. The engine parameters are summarized in Table 4.3. The measuring points are the same as for the ignition time experiment.

Engine parameter	Value	Engine parameter	Value
n_{engine}	8000 rpm	$T_{\text{o,i}}$	387.15 K
t_{int}	301.15 K	$T_{\text{w,i}}$	386.5 K
t_{amb}	294.65 K	m_{fuel}	27 - 52 mg/stroke
λ_{cmb}	1.21	α_{ign}	37.5 - 44.0 °CA

Table 3.3: Engine parameters during the variation of boost pressure at a constant air-fuel ratio.

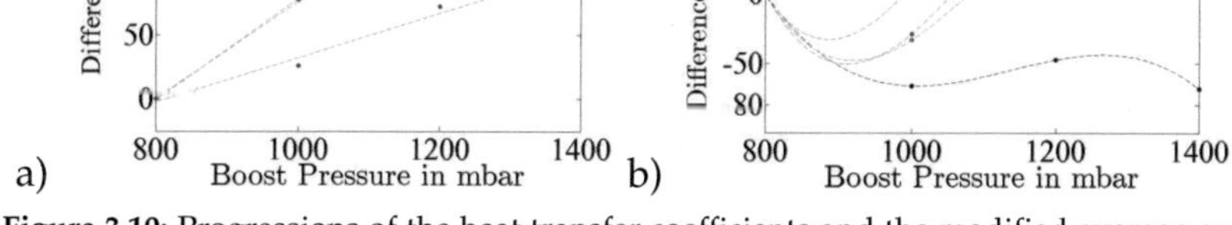

Figure 3.10: Progressions of the heat transfer coefficients and the modified average gas temperatures: a) HTC increase in the combustion chamber due to more boost pressure for three different models. b) Corresponding ACT development with increasing boost pressure. For comparative purposes, the average temperature (not modified) is plotted.

In Fig. 3.10, the variations of the heat transfer coefficients and the modified average gas temperatures are shown. The dashed lines are cubic and linear regression lines. For comparative purposes, the average gas temperature (not modified) is plotted. It is just the resulting temperature after the ideal gas law. Of course, it is the same for every model. The difference results from the non-linearity of Eq. (3.3) and reflects the great importance of correlations like Eq. (3.6). In Fig. 3.11, the simulation and experimental results are shown for the solid temperatures. Again, dashed lines are added for a better evaluation of sensitivities. The results are comparable with Choi et al. (1997): increasing load also resulted in an approximately linear temperature increase in the same order of magnitude. Given the small disparities between simulation and experiment in Fig. 3.11 a), it is a remarkable result that every model can predict sensitivities to boost pressure variations within a small range. Eichelberg's model is not as effective as the two other. The models according to Woschni and Hohenberg are within the measurement uncertainty of the thermocouples.

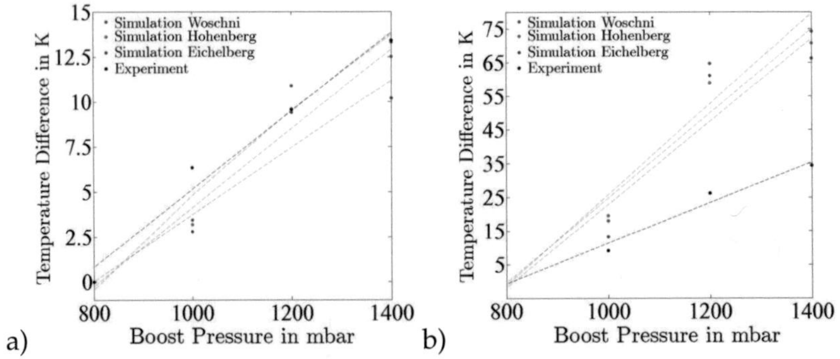

Figure 3.11: Comparison between simulation and experimental results for boost pressure variation: a) temperature increase at the combustion chamber wall due to higher loads. b) Corresponding temperature rise at the spark plug with increasing boost pressure.

Beyond the throttled condition, the temperature difference at the combustion chamber wall is in good agreement with the experimental results.

Besides a variation of the induced torque, boost pressure variations with a constant air-fuel ratio result in a continuous change in the amount of gas within the combustion chamber. That is the reason for this kind of experiment to be more challenging than pure ignition time variations. The progress of the modified average gas temperature is much more complicated, see Fig. 3.10 b). The striking similarity between the models according to Woschni and Hohenberg, in this case, can be seen in Figs. 3.10 a) and 3.11 a). The differences between the average heat transfer coefficients and the resulting temperatures at the combustion chamber wall are quite small. For load variations at a constant engine speed, this behavior can also be found in Wimmer et al. (2000): plotting the isolines for the time averaged heat transfer ratio between the models according to Woschni and Hohenberg within a load-speed diagram, both models behaves quite similar with increasing engine load. This means that the isolines are almost perpendicular to the engine speed axis. Other investigated models like Bargende (Chiodi and Bargende, 2001) or Kleinschmidt (Kleinschmidt, 1993) show larger modifications. The model according to Bargende includes the turbulent kinetic energy and describes the combustion process with the help of a two zone model. Kleinschmidt uses partial differential equations for the description of physical processes within a combustion chamber. The lower gradient of the heat transfer coefficient according to Eichelberg's model is balanced with the higher sensitivity to the modified average gas temperature resulting in similar solid temperature predictions. As can be seen in Fig. 3.11 b), each model overpredicts significantly the spark plug temperatures with increasing boost pressure. At the highest boost pressure of 1400 mbar, the temperature difference between experiment and simulation is equivalent to the measured temperature increase. It seems that the simplified spark plug modeling approach according to Eq. (3.8) overestimates the higher heat transfer to spark plugs. A more reasonable approach should not only account for the additional heat, resulting from another position relative to the combustion chamber, but also from the ignition process itself. This would decrease the sensitivity to changes within the combustion process. According to Eq. (3.2), a new simplified approach could be:

$$\boldsymbol{q} = -\alpha_{\text{increase}}\alpha_{\pi}\left(T_{ref} - T_{\text{spark}}\right)\boldsymbol{n} - \boldsymbol{q}_{\text{Ign}}. \tag{3.10}$$

The nomenclature is the same as in Eq. (3.8). $\boldsymbol{q}_{\text{Ign}}$ is the additional heat flux vector due to the ignition process. Its offset function reduces the sensitivity to changes within the combustion process. However, this vector is probably a quite complicated function of the local thermodynamical state around the spark plug and the required ignition voltage. Since modeling of such a formation of free radicals is not that easy and the spark plug is not in the main focus of this thesis, Eq. (3.10) will not be further investigated.

3.5.4 Lambda variation by adjusting boost pressure

While keeping the engine speed and the fuel mass flow constant, the boost pressure is varied resulting in different values for the air-fuel ratio. The ignition time is adjusted manually to a specific thermal efficiency and knocking level. Besides the variation of the heat transfer coefficients in the combustion chamber and the modified average gas temperature, the differences in the boost pressure at a constant fuel mass flow rate have another effect which results in a variation of the heat transfer coefficients in the outlet channel and the outlet valve stem. In addition, the exhaust temperature varies because of different thermal efficiencies. Moreover, the heat capacity of the exhaust gas varies with different air-fuel ratios. All in all, an experiment with a variation of the mixture is a problem which contains lots of different aspects: changes in the thermochemical status in the combustion chamber, in the absolute number of molecules and in the pressure curves during the working cycle. The differences in the exhaust heat transfer coefficients and temperatures come along. In Table 4.2, the engine parameters are summarized.

Engine parameter	Value	Engine parameter	Value
n_{engine}	8000 rpm	$T_{\text{o,i}}$	364.15 K
t_{int}	306.15 K	$T_{\text{w,i}}$	356.5 K
t_{amb}	300.15 K	m_{fuel}	91.25 mg/stroke
λ_{cmb}	0.9 - 1.4	α_{ign}	26.0 - 42 °CA

Table 3.4: Engine parameters during variation of air-fuel ratio at a constant mass flow rate

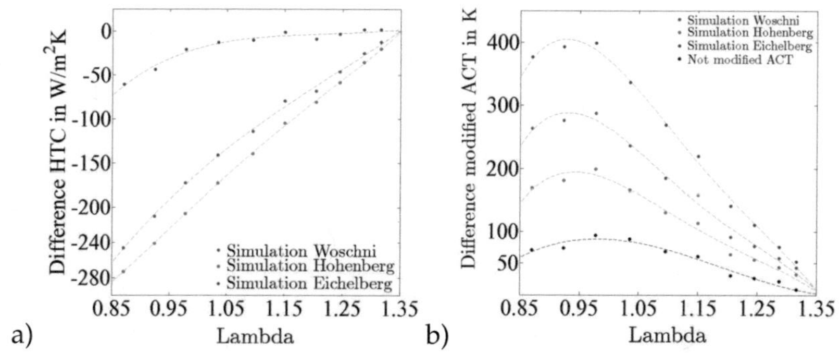

Figure 3.12: Progressions of the heat transfer coefficients and the modified average gas temperatures as a function of air-fuel ratio. The dashed lines are fitted polynomials of the fourth order using the method of least squares: a) HTC increase in the combustion chamber due to higher air-fuel ratios for three different models. b) Corresponding ACT development with increasing air-fuel ratios. For comparative purposes, the average temperature (not modified) is plotted.

In Fig. 3.12, the resulting differences in the heat transfer coefficients and modified average gas temperatures for each model are presented. The highest value for λ_{cmb} serves as reference point. Again, one can see large differences between the Eichelberg model and the two other models. The tendency is the same as in Fig. 3.10. Remember that now the highest value for the air-fuel ratio is set as a reference point. According to Fig. 3.12 a), the Eichelberg model predicts little changes in the modified average gas temperature during λ_{cmb} - variations. In addition, this model shows the highest increase in the modified average gas temperature with increasing air-fuel ratio. Both result in a complete overestimation of solid temperature differences, as can be seen in Fig. 3.13 a). The two other models behave quite similar concerning the development of the heat transfer coefficient in the combustion chamber. Hohenberg's prediction for the average gas temperature is, however, too weak, resulting in an underestimation of solid temperatures, as can be seen in Fig. 3.13 a). The prediction according to Woschni is the best compared to the other two models which is the result of interaction between the change in the heat transfer coefficient and the average gas temperature. While the first falls continuously with increasing fuel-air ratio, the second has a maximum in rich combustion conditions. This example of different air-fuel ratios shows clear differences in the models under investigation. In Fig. 3.12, the average gas temperature (not modified) is plotted. Again, one can see the large importance of the correlation term (3.6). The modified gas temperature is clearly larger than the average gas temperature resulting in a higher heat flux to the wall. This means that the correlation must be positive: $\langle \alpha' T'_{ref} \rangle \geq 0$. This is physically meaningful, because during the combustion stroke, the average gas temperature and the heat transfer coefficient have its maximum shortly after top dead center. In addition, both maximum values show a particularly strong amplitude with an enormous contribution to the correlation term. In Fig. 3.12 b), one can also see that the physical average gas temperature, e.g., not modified, has its maximum at $\lambda_{\mathrm{cmb}} \approx 1$. Actually, it is a bit lower than one. This is also physically meaningful, because it is well known that the laminar flame speed and the adiabatic flame temperature have their maximum values when λ_{cmb} is a little bit lower than one (Warnatz et al., 2006). This is also the reason why the maximum value of the exhaust ring temperature occurs at the stoichiometric region (see Fig. 3.13 a)).

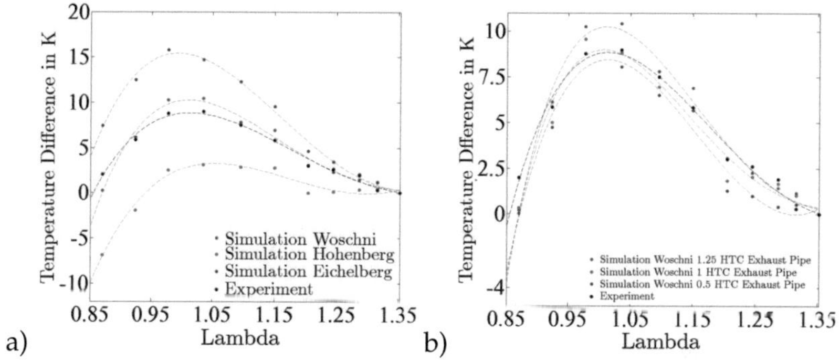

Figure 3.13: Progression of the exhaust valve seat temperature as a function of the air-fuel ratio. The highest value for λ_{cmb} is set as a reference point. The dashed lines are fitted polynomials of the fourth order using the method of least squares: a) simulated and measured temperatures for three different models. b) Simulated and measured temperatures for the Woschni model with three different exhaust pipe heat transfer coefficients.

Another interesting aspect can be seen in Fig. 3.13 b). For comparison reasons, simulations have been performed with different heat transfer coefficients in the exhaust pipe. Therefore, a slightly larger and lower HTC value is investigated. Of course, the heat transfer coefficient in the outlet pipe has an influence on solids which are located next to the outlet. Lower values result in a less sensitivity to air-fuel ratios. This can be explained with the influence of the exhaust gas temperature which is a complicated function of thermal efficiency and effective heat capacity of the exhaust gas mixture. The progression of the exhaust gas temperature is qualitatively similar to the temperature curve in Fig. 3.13 b).

3.6 Conclusions

With regard to component temperatures and their dependence on different engine settings, a simplified, statistical CFD-CHT calculation method is presented. Concerning the initial research question, if such a method without simulating crank angle resolved heat fluxes is purposeful for determining cycle averaged component temperatures, the result is as follows.

In spite of the difference concerning the instantaneous heat flux, discussed in numerous publications, the proposed simulation technique, which uses integrated boundary conditions, together with the Buckingham Pi-Theorem, more precisely, with simplified engine specific correlations, is in general able to predict relative differences in cycle averaged component temperatures as a function of the engine setting. If the calculation time is an important factor, it is appropriate to use these simplified models. One has, however, to pay attention which model in detail is used. The Woschni model slightly overpredicts the increased heat transfer with increasing ignition time, whereas the other two models underpredict the effect. The differences to measurements are in the range of 2 K. This is small compared to a measurement inaccuracy of 1 K. For boost pressure variations, all models describe a similar sensitivity to load changes even if individual temperature differences up to 4 K can be identified. In contrast, the correlation according to Woschni is clearly the best choice for Lambda variations: the other two models incorrectly calculate the maximum temperature by the amount of 8 K, whereas the model according to Woschni is within the range of the measurement error.

Another main conclusion is the importance of significant non-linearity effects in heat transfer problems when engine cycles are not simulated in detail. Concerning gas temperatures, the difference between the pure expectation value and the statistically modified average value, which takes the correlation between the heat transfer coefficient and the gas temperature into account, can be up to 300 K, depending on the model.

Chapter 4

Fast algorithms for generating thermal boundary conditions in combustion chambers[1]

4.1 Outline of the chapter

The research question can be formulated as follows: following Chapter 3, concerning crank angle averaged solid temperatures, is it possible to identify accurate enough thermal boundary conditions only with cycle integrated quantities as model input variables? In particular, for the purposes of engineering applications, can the conflict of objective between the calculation time, model input variables and the sensitivity to diverse engine settings be solved with an expedient modeling?

Starting from a reference operating point, for which heat transfer coefficients and the average gas temperature are known, relative changes in these quantities are calculated. Using pressure indication measurements and the model according to Woschni (1967), thermal boundary conditions for this reference state are determined. As a function of cycle integrated quantities, like induced engine torque or fuel mass flow, relative changes in these boundary conditions are calculated. Probability density functions and realisations of thermal quantities, which

[1] This chapter is based on the paper "Fast algorithms for generating thermal boundary conditions in combustion chambers" (Hölz et al., 2018b) **DOI: https://doi.org/10.1016/j.applthermaleng.2018.05.099**

are now interpretated as random variables, are, therefore, transformed according to different thermodynamical operating settings. In this context, variations in ignition time and air-fuel ratio as well as boost pressure are investigated. To compare the experimentally determined temperatures with the proposed method, CFD-CHT simulations with the proposed thermal boundary conditions are performed. Again, the engine water jacket is simulated with the SST k-ω turbulence model from Chapter 2. Taking the example of ignition time variations, heat flux increases, resulting from highly non-uniform pressure changes over crank angle and the non-linear nature of heat transfer, are discussed and the corresponding limits for such cycle integrated methods are presented in detail.

4.2 Heat transfer in the combustion chamber - variation of the thermodynamic states -

Assuming forced convection according to Eq. (3.3), all fluctuations and higher frequencies, resulting from the oscillating function of engines, are filtered out. Interpreting heat transfer coefficients and gas temperatures during statistically enough engine cycle as random variables, one has for any arbitrary random variable $\tilde{f}(\alpha)$ Eq. (5.3) for its expectation value. It can be used correspondingly for the reference temperature T_{ref} or, according to Eq. (3.3), the product (αT_{ref}).
An overview of the approach is shown in Fig. 5.5. Given are the probability density functions at the condition $\underline{M}_1$, derived with measured pressure data from test bench under stationary conditions. The engine state is characterized with the help of Eq. (3.1). In dependence of the state of interest $\underline{M}_2$, which should have the same engine speed as $\underline{M}_1$, the new probability density functions for the heat transfer coefficients and the corresponding reference temperatures are modeled. In this chapter,

all variances from stationary reference conditions at test bench are called *part load* conditions. Reference states are called *full load* conditions.

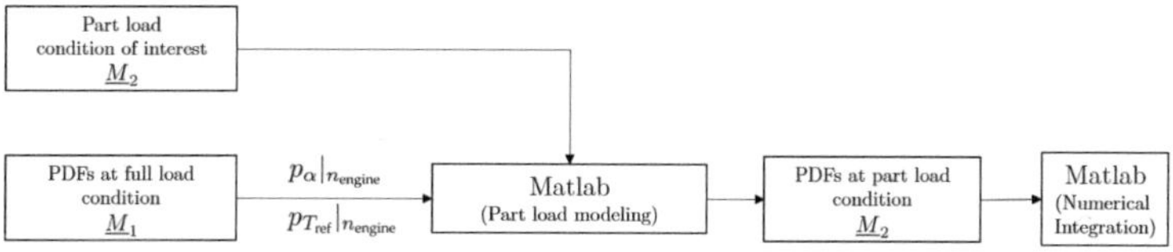

Figure 4.1: Method overview for generating probability density functions for HTC and ACT at part load condition $\underline{M}_2$. Given are the probability density functions at full load condition $\underline{M}_1$. $\underline{M}_1$ and $\underline{M}_2$ should have the same engine speed.

The idea is to model the probability density functions at part load conditions with the help of general transformation rules. Given is a random variable U which has the PDF $f_u(u)$. Suppose there exists a transformation rule Q between U and another random variable $\tilde{U}$ with PDF $f_{\tilde{u}}(\tilde{u})$:

$$\tilde{U} = Q(U). \tag{4.1}$$

The corresponding realisations of the random variables U and $\tilde{U}$ are given by u and $\tilde{u}$. Then, the PDF $f_{\tilde{u}}(\tilde{u})$ is given by:

$$f_{\tilde{u}}(\tilde{u}) = f_u(u) / \left| \frac{\mathrm{d}Q(u)}{\mathrm{d}u} \right|, \tag{4.2}$$

where the transformation between realisations is defined by

$$\tilde{u} \equiv Q(u). \tag{4.3}$$

To model PDFs for part load conditions, one has to define the transformation Q or, more specifically, its derivative $\mathrm{d}Q(u)/\mathrm{d}u$, for the engine state of interest $\underline{M}_2$. So a transformation of typ

$$Q = Q(u)|_{\underline{M}_2} \tag{4.4}$$

is searched. The method is described schematically in Fig. 4.2.

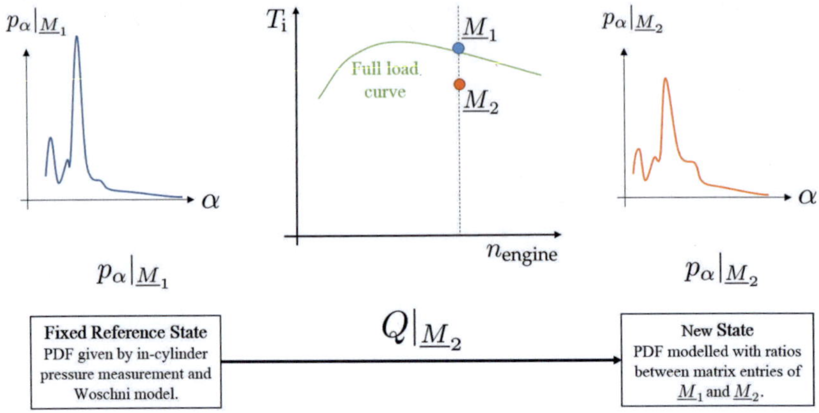

Figure 4.2: Schematic representation: transformation between probability density functions $p_{\alpha}|_{\underline{M}_1}$ and $p_{\alpha}|_{\underline{M}_2}$. For a given reference state $\underline{M}_1$, the new PDF for the state $\underline{M}_2$ is calculated with a transformation $Q|_{\underline{M}_2}$.

According to Eq. (5.3), it follows for the heat transfer coefficient α in the new engine state $\underline{M}_2$:

$$\langle \alpha \rangle = \int\limits_{\mathbb{R}_{\geq 0}} A p_{\alpha|\underline{M}_2}(A)\, dA. \tag{4.5}$$

In this case, the function $\tilde{f}$ is the identity function. In the following, a model for the transformation $Q|_{\underline{M}_2}$ is suggested. Let α_1 and α_2 be two realisations for the heat transfer coefficient in the state $\underline{M}_1$ and $\underline{M}_2$, respectively. A simple approach for the transformation is

$$Q|_{\underline{M}_2}(\alpha_1) = \beta_T(\underline{M}_1, \underline{M}_2, \alpha_1)\alpha_1, \tag{4.6}$$

with a transformation coefficient $\beta_T(\underline{M}_1, \underline{M}_2, \alpha_1)$ which, in general, also depends on the realisation α_1. An explanation follows below. According to Eq. (4.3) and (4.6), the realisation α_2 is given by

$$\alpha_2 = \beta_T(\underline{M}_1, \underline{M}_2, \alpha_1)\alpha_1, \tag{4.7}$$

with the ratio between the two realisations

$$\beta_T(\underline{M}_1, \underline{M}_2, \alpha_1) = \alpha_2/\alpha_1. \tag{4.8}$$

Starting from the correlation by Woschni (Woschni, 1967)

$$\alpha \propto p^m v^m \overline{T}_g^{0.75-1,62m}, \tag{4.9}$$

with a Reynolds exponent of $m = 0.8$, one can get the transformation $Q(u)|_{\underline{M}_2}$ if one models the ratios of gas pressure p, characteristic velocity v, and in-cylinder gas temperature $\overline{T}_g$ between the part load and the stationary full load condition. Assuming a relation according to Eq. (4.9), the transformation coefficient β_T can be described by

$$\beta_T(\underline{M}_1, \underline{M}_2, \alpha_1) = \left(\frac{p|_{\underline{M}_2}}{p|_{\underline{M}_1}}\right)^m \left(\frac{v|_{\underline{M}_2}}{v|_{\underline{M}_1}}\right)^m \left(\frac{\overline{T}_g|_{\underline{M}_2}}{\overline{T}_g|_{\underline{M}_1}}\right)^{0.75-1.62m}. \tag{4.10}$$

The original models according to Woschni and Hohenberg determines the gas temperature $\overline{T}_g$ with the ideal gas law: $\overline{T}_g = pV/NR$. This gas temperature is used as reference temperature in Eq. (3.3). Remember the characteristic velocity according to Woschni:

$$v = v_{\text{c1}} + v_{\text{c2}} \tag{4.11}$$

$$= c_1 v_{\text{piston}} + c_2 V_{\text{d}} \left(\frac{\overline{T}_g}{pV}\right)_{\text{IVC}} (p - p_m). \tag{4.12}$$

Because the characteristic velocity consists of two summands v_{c1} and v_{c2}, the ratio $(v|_{\underline{M}_2})/(v|_{\underline{M}_1})$ in Eq. (4.10) has to be a function of the realisation α_1. For the complete pressure $p = p_{\text{cmb}} + p_{\text{m}}$ in Eq. (4.10), it is the same difficulty: the corresponding weighting factor for each summand is a function of the realisation α_1. In Fig. 4.3 and 4.4, the weighting factors for the combustion pressure p_{cmb} and the velocity part v_{c2} are shown for different engine speeds. The data result from in-cylinder pressure measurements. For engines, it is worthwhile to divide a cycle into four strokes. Afterwards, one has to integrate the resulting PDFs separately (Eq. (5.3)). To calculate the complete number of molecules N_{com} before and after the chemical reaction, following simplified equation is assumed:

$$\mathrm{C_xH_yO_z} + (\mathrm{x} + \mathrm{y}/_4 - \mathrm{z}/_2)\mathrm{O_2} \longrightarrow \mathrm{xCO_2} + \mathrm{y}/_2\mathrm{H_2O}.$$

For over-stoichiometric conditions, the difference between the actual number and the stoichiometric number of O_2 molecules is added on the product side. Accordingly, it is assumed that in lean combustion not required oxygen in excess behaves like an inertgas. For under-stoichiometric conditions, there is no oxygen in the exhaust gas and the following water-gas shift reaction is assumed (Merker and Teichmann, 2014), resulting in no change in the number of molecules:

$$\mathrm{CO_2} + \mathrm{H_2} \rightleftharpoons \mathrm{CO} + \mathrm{H_2O}.$$

An approximation for the, in general, temperature dependent equilibrium constant can be found in JANAF (1971). In the following, some approximations for the specific ratios in Eq. (4.10), only as functions of the ratios between the matrix entries $\underline{M}_2$ and $\underline{M}_1$, are suggested. For a simplified notation, only the relation of proportionality is given. The subscripts *before* and *after* refer to the combustion state. Assuming an homogeneous operation during the intake stroke, the first and second stroke is given by:

$$\begin{aligned} p &\propto t_{\text{int}}(N_{\text{com}})_{\text{before}}, \\ \overline{T}_g &\propto t_{\text{int}}, \\ v &= const. \end{aligned}$$

Third stroke (v_{c2} is the second summand of Eq. (4.12)):

$$\begin{aligned} p_{\mathrm{m}} &\propto t_{\mathrm{int}}(N_{\mathrm{com}})_{\mathrm{before}}, \\ p_{\mathrm{cmb}} &\propto T_{\mathrm{i}}, \\ \overline{T}_g &\propto 2p/((N_{\mathrm{com}})_{\mathrm{before}} + (N_{\mathrm{com}})_{\mathrm{after}}), \\ v_{c2} &\propto T_{\mathrm{i}}/(N_{\mathrm{com}})_{\mathrm{before}}. \end{aligned}$$

Fourth stroke ($\Delta\overline{T}_g$ corresponds to increase of gas temperature based on inlet temperature due to combustion):

$$\begin{aligned} \Delta\overline{T}_g &\propto \sqrt{\frac{m_{\mathrm{fuel}}}{T_{\mathrm{i}}} \frac{m_{\mathrm{fuel}}}{(N_{\mathrm{com}})_{\mathrm{after}}}}, \\ p &\propto \overline{T}_g(N_{\mathrm{com}})_{\mathrm{after}}, \\ v &= const. \end{aligned}$$

As an example, the amount of substance $(N_{\mathrm{com}})_{\mathrm{before}}$, or $(N_{\mathrm{com}})_{\mathrm{after}}$ respectively, could also include the EGR (Exhaust Gas Recirculation) fraction, which serves as an additional inert gas. In this case, it should be noted that the inlet gas temperature has to be modified according to the mixing temperature of the exhaust and fresh gas. The first two strokes are quite simple. Approximately, only the amount of gas and the inlet temperature are significant. Assuming a correlation of the form pV^{κ} = const. in the second stroke, with κ as the isentropic exponent, an adiabatic and reversible compression is modeled. Because only the first summand of Eq. (4.12) is relevant, the characteristic velocity keeps constant. In the third stroke, it is assumed that the induced torque T_{i} is mainly influenced by the combustion pressure p_{cmb} during expansion. Therefore, the gas exchange period is neglected. Again, it should be noted that the cylinder pressure consists of two parts: the motored part and the part resulting from the combustion. In this very simplified model, the combustion part is proportional to the combustion torque. The motored pressure is treated analogously to the second stroke. The characteristic velocity in the third stroke is treated in a similar way. Only the second part of the Woschni velocity (4.12) is changed. As already mentioned, the weighting factors for the two combustion parts p_{cmb} and v_{c2} are, therefore, necessary. Fig. 4.3 and 4.4 show these ratios as function of crank angle and in the HTC space during the expansion stroke.

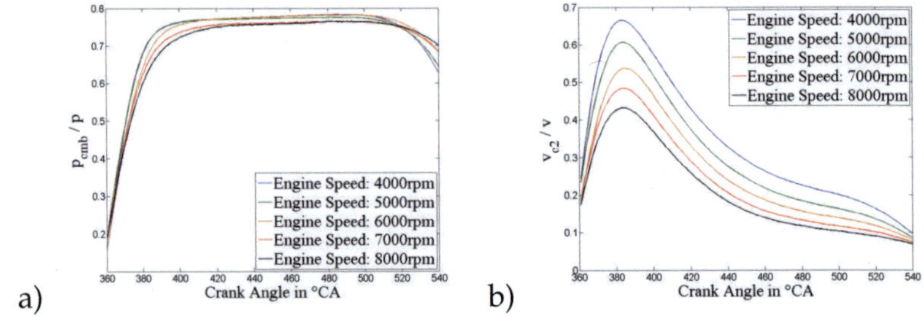

Figure 4.3: a) Ratio p_{cmb}/p over crank angle for different engine speeds. b) Ratio v_{c2}/v in Eq. (4.11) over crank angle for different engine speeds.

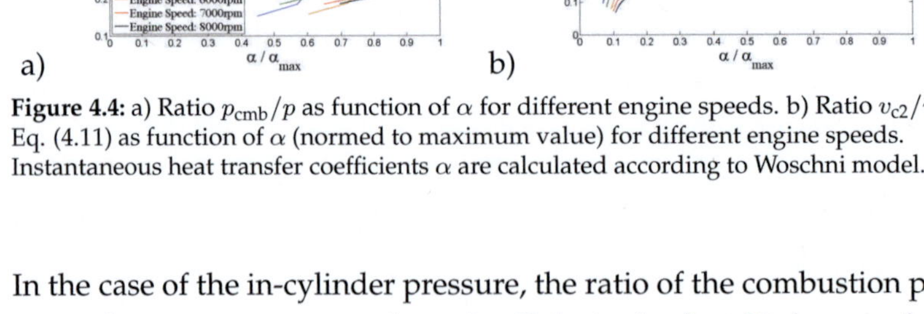

Figure 4.4: a) Ratio p_{cmb}/p as function of α for different engine speeds. b) Ratio v_{c2}/v in Eq. (4.11) as function of α (normed to maximum value) for different engine speeds. Instantaneous heat transfer coefficients α are calculated according to Woschni model.

In the case of the in-cylinder pressure, the ratio of the combustion part is nearly constant over crank angle. Only in the first 20 degrees there is a significant increase. In this chapter, a constant value is assumed.

In the case of the characteristic velocity, it is more complicated. As can be seen in Fig. 4.3 b) and 4.4 b), there is a large dependence on time and HTC. According to modeling depth, the dependence on the heat transfer coefficient has to be modeled. Remember that the complete transformation process is performed in the HTC space. That is the reason why the ratios p_{cmb}/p and v_{c2}/v have to be described as functions of α. In turn, the transformation coefficient β_T from Eq. (4.10) also depends on α. In this chapter, a linear correlation between the minimum and maximum value is used. In general, as can be seen in Fig. 4.3 and 4.4, the weighting factors should be defined as functions of engine speed.
In the last stroke, the temperature difference due to combustion is modeled. The heat capacity of the gas is, therefore, assumed to be independent of temperature, pressure and gas composition. The ratio of m_{fuel} and T_{i} is a measurement of efficiency: multiplied with the mass flow of fuel it gives a ratio of energy. According to Heywood (1988), approximately one-half of the energy which does not contribute to the indicated power results in the exhaust enthalpy flux entering atmosphere. With respect to this approximation, the square root is used for calculating the temperature difference $\Delta\overline{T}_g$. For small variations from the stationary full load state, a Taylor expansion gives for the square root function $f(x) = \sqrt{x}$:

$$f(x + \Delta x)|_{x=1} \approx 1 + \Delta x/2. \tag{4.13}$$

Comments to the proposed model:

Due to the unique allocation (4.1), all statistical properties are transformed accordingly $Q(u)$. However, it is clear that, due to changes in load conditions, the scattering nature of the system will change. Furthermore, the modeling of part load conditions in rich combustion is more complex than in lean combustion. In addition to imperfect combustion due to chemical kinetics, there is the effect of an incomplete reaction due to the lack of oxygen: chemical reaction-pathways and equilibrium states are different. Defining a degree of implementation $\eta_u = 1 - Q_{\text{inc}}/\left(m_{\text{f}} H_u\right)$ with the non-released energy Q_{inc}, the fuel mass m_{f} and the calorific value H_u of the fuel, one can describe this effect. As can be seen in Fig. 4.5, the degree of implementation drops very fast with lower values of λ_{cmb}.

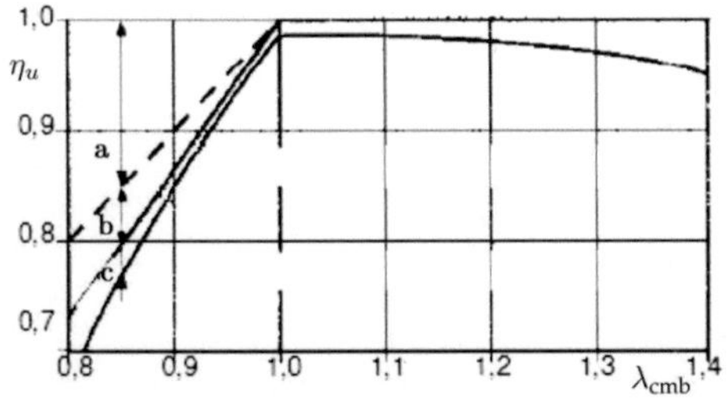

Figure 4.5: Degree of implementation η_u as a function of λ_{cmb}. a: Loss due to oxygen lack; b: Loss due to incomplete combustion; c: Loss due to imperfect combustion (Eifler et al., 2009).

One simplified possibility to account for this drop is the consideration of the λ_{cmb} ratio between the stationary full load and part load condition. Defining a modified ramp function f_r for positive real-valued input arguments λ,

$$f_r(\lambda) = \begin{cases} (6/9)\,(\lambda - 1) + \lambda, & \lambda \leq 1 \\ 1, & \lambda > 1 \end{cases}$$

one can modify the temperature difference due to combustion with the ratio between part and full load condition:

$$\Delta \overline{T}_g \propto \frac{f_r(\lambda_{\text{cmb}})|_{\text{Part load}}}{f_r(\lambda_{\text{cmb}})|_{\text{Full load}}}. \tag{4.14}$$

An additional aspect is the influence of the evaporation enthalpy of the injected fuel. One possible way to take this effect into account is the modification of the inlet temperature. The ratio between fuel mass flow rate and the total mass flow rate is, therefore, essential:

$$\Delta t_{\mathrm{int}} = -\frac{c_{\mathrm{evaporation}} m_{\mathrm{fuel}}}{c_{\mathrm{mix}} \left(m_{\mathrm{fuel}} + m_{\mathrm{air}} \right)}. \tag{4.15}$$

The evaporation enthalpy is given by $c_{\mathrm{evaporation}}$, whereas the effective heat capacity of the mixed gas is described by c_{mix}. In the Appendix A, by the example of a pseudo code for the HTC, a possible way of programming is shown.

4.3 Results and discussion

Temperature measurements are used for calibration and validation of the above presented model. Therefore, a 0.5 liter turbocharged single cylinder SI engine is investigated under varying boundary conditions. For stationary tests, special focus is placed on specific variations of the engine mapping, including ignition time, air-fuel ratio and boost pressure. Type K Thermocouples, which have an uncertainty of 1 K, are used for the measurement of solid temperatures. Heat-conducting paste with a thermal conductivity of 2.5 W/mK is put between the solid surface and the thermocouples. Measuring points are installed inside of the spark plug coat, at the combustion chamber wall of the cylinder head and the exhaust valve seat (seat ring). Therefore, a thermocouple is welded into the spark plug which ended at the last thread. The position in the combustion chamber is located at the outer diameter, 1 mm above the surface. With a sampling rate of 0.1 °CA, 60 combustion cycles for each measuring point are recorded with the help of a piezo-quartz pressure transducer: pressure curves are needed for comparison purposes with the Woschni (Woschni, 1967) and Hohenberg model (Hohenberg, 1979). Fig. 4.6 gives an overview about the experimental validation method.

The simulated solid temperature difference between the states $\underline{M}_1$ and $\underline{M}_2$, which is based on the thermal boundary conditions of the suggested calculation algorithm, is compared with the measured temperature difference.

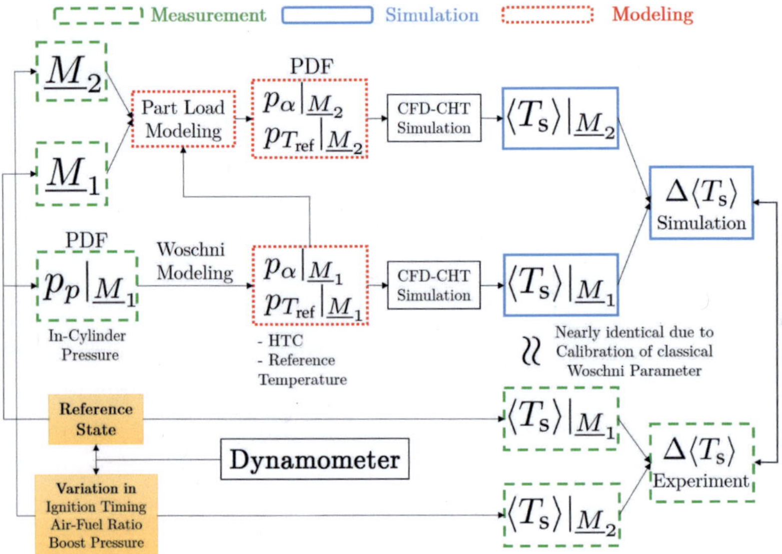

Figure 4.6: Overview about the experimental validation method. The probability density functions for the heat transfer coefficient p_α and the reference temperature $p_{T_{\text{ref}}}$ (in this case the average in-cylinder gas temperature) serve as thermal boundary conditions for subsequent CFD-CHT simulations.

4.3.1 Variation of ignition time

In Fig. 4.7, the differences in the heat transfer coefficient and modified average gas temperature for an ignition time variation are shown. The engine operation state is summarized in Table 4.1. Interestingly, the suggested part load model predicts values for the heat transfer coefficient between the Woschni and the Hohenberg model. The average gas temperatures are slightly lower than the values calculated by the Hohenberg model. The lowest ignition time is chosen as a reference point for the part load model. In Fig. 4.8, differences in the cylinder head temperature are shown.

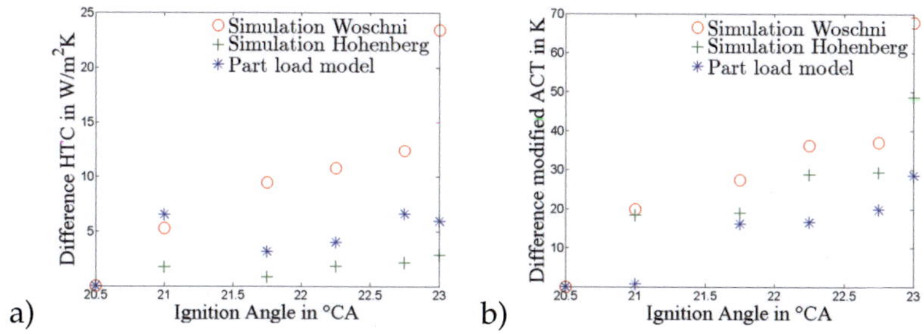

Figure 4.7: Progressions of the heat transfer coefficients and the modified average gas temperatures as function of ignition time. a) HTC increase in the combustion chamber due to higher ignition angles for three different models. b) Corresponding ACT development with increasing ignition angle.

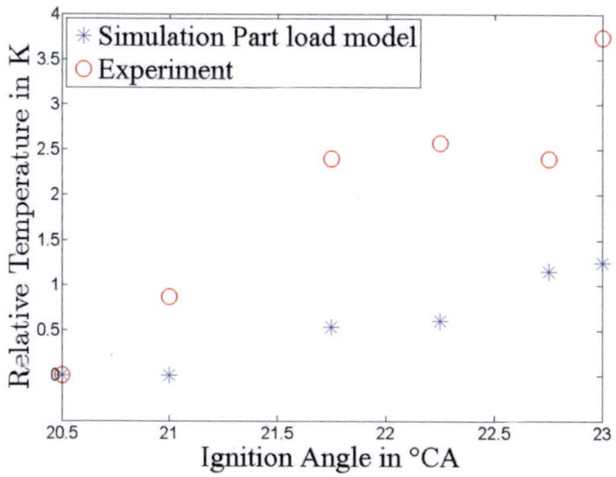

Figure 4.8: Progression of cylinder head temperature as function of ignition time.

Parameter	Value	Parameter	Value
n_{engine}	6000 rpm	$T_{\text{o,i}}$	388.15 K
t_{int}	312.15 K	$T_{\text{w,i}}$	387.5 K
t_{amb}	294.15 K	m_{fuel}	115.5 mg/stroke
λ_{cmb}	1.2	α_{ign}	20.5 - 25.25 °CA

Table 4.1: Engine parameters during variation of ignition angle.

As can be seen, the part load model underestimates the sensitivity to ignition time variations. In this case, the assumption of a constant pressure ratio in the expansion stroke $p_{\text{cmb}} \propto T_{\text{i}}$ is a strong simplification. As can be seen in Fig. 4.9, the ratio is highly non-uniform.

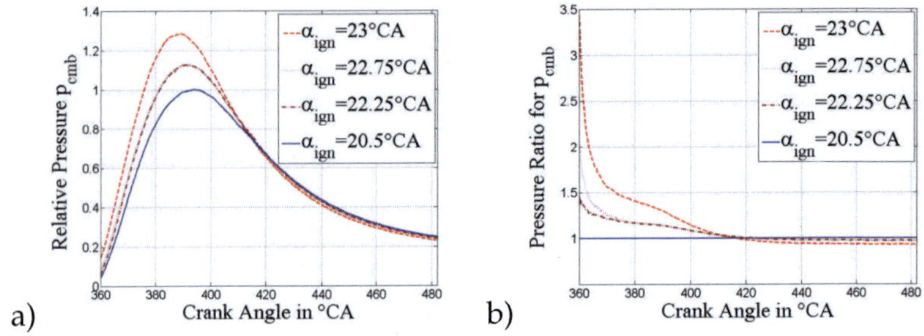

Figure 4.9: Pressure ratios over crank angle. a) Pressure due to combustion p_{cmb}. Values are normalized to the highest pressure value of the lowest ignition angle. b) Pressure due to combustion p_{cmb}. Values are instantaneously normalized to the pressure curve of lowest ignition angle.

There are mainly three reasons for the discrepancies in Fig. 4.8:

i) The localised, non-uniform pressure ratio increase: the non-linear dependence of pressure in Eq. (4.9) causes an increased heat flux.

ii) The engine kinematics: For a given in-cylinder gas pressure, the tangential force at the crankshaft is a function of crank angle α_{cr}. Therefore, for a given torque increment $\mathrm{d}T_{\text{i}}$ and neglected rod ratio, it follows:

$$dT_\mathrm{i} \propto dp_\mathrm{cmb} \sin\left(\alpha_\mathrm{cr}\right). \tag{4.16}$$

For the same torque dT_i, pressure jumps dp_cmb near top dead center need higher pressure ratios resulting in an increased heat flux. However, as can be seen in Fig. 4.3, this effect is slightly weakened by the fact that the ratio of the combustion part for pressure and the characteristic velocity is lower in beginning of the expansion stroke.

iii) A pressure increase before top dead center is counterproductive. Such amounts have to be compensated in the following expansion stroke which results in higher heat fluxes.

All three effects are illustrated qualitatively in Fig. 4.10. In the case of the presented ignition time variation, the effect of localisation on solid temperatures is investigated in Fig. 4.11. One can see that modeling according to a jumping function improves the reliability. However, any other model simplifications, including the above-mentioned effects ii) and iii), cause further differences. For the jumping function modeling, heat flux ratios are calculated according to the simplified correlation for the heat flux:

$$q \propto \alpha \overline{T}_g \propto p^m v^m \overline{T}_g^{0.75-1.62m} \overline{T}_g \propto p^m v^m p^{0.75-1.62m} p = v^m p^{1.75-0.62m}. \tag{4.17}$$

Compared to the hot gas temperatures during the expansion stroke, the first approximation neglects solid temperatures. In the second derivation, the Woschni model is used (see Eq. (4.9)). The third simplification results from the ideal gas law: $\overline{T}_g \propto p$ for a constant amount of molecules. Compared to the part load model, Eq. (4.17) is used to calculate the heat flux ratio for the square function: The width of the rectangle is assumed to be 30 °CA. The pressure height results from the requirement for a constant torque.

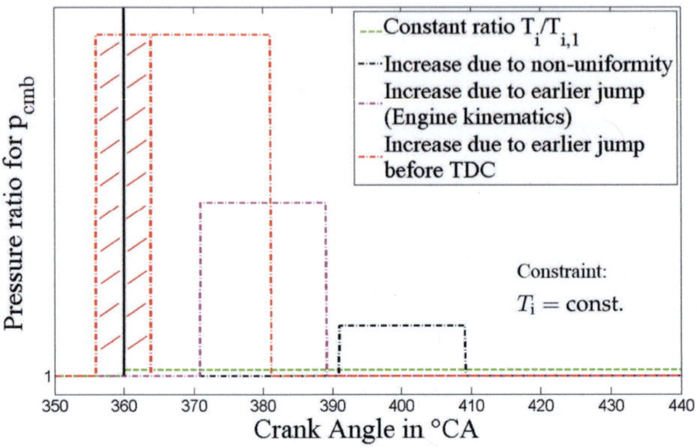

Figure 4.10: Different necessary pressure ratios for a given torque increase $T_i/T_{i,1}$. $T_{i,1}$ corresponds to the engine state $\underline{M}_1$.

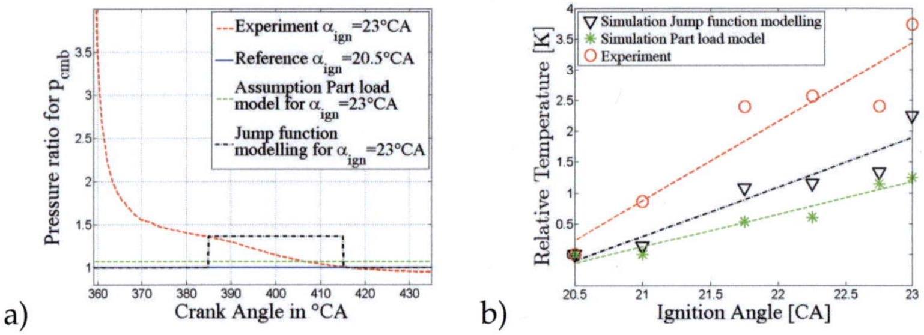

Figure 4.11: a) Pressure ratios over crank angle. Pressure due to combustion p_{cmb} is instantaneously normalized to the pressure of lowest ignition angle. All three curves for an ignition angle of 23 °CA result in the same torque increase. b) Corresponding cylinder head temperatures as function of ignition angle.

4.3.2 Variation of air-fuel ratio

In Fig. 4.12, the differences in the heat transfer coefficient and the modified average gas temperature for various air-fuel ratios are shown. The engine operation state is shown in Table 4.2. The highest air-fuel ratio is chosen as a reference point for the part load model. Compared with the Woschni model, quite similar results for the heat transfer coefficients can be achieved. Extremely small deviations can be observed in Fig. 4.12 a).

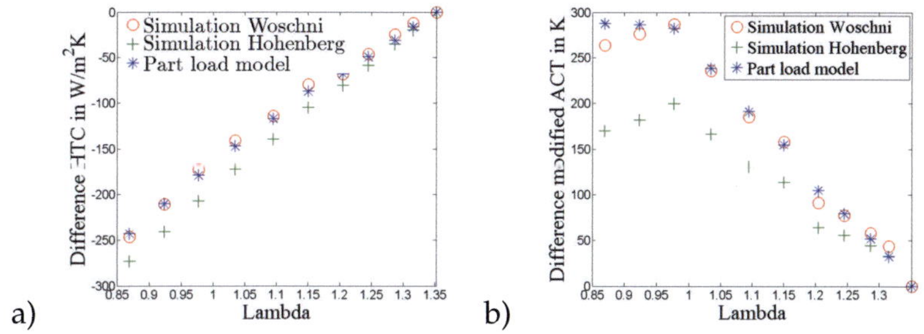

Figure 4.12: Progressions of the heat transfer coefficients and the modified average gas temperatures as function of air-fuel ratio. a) HTC increase in the combustion chamber due to higher air-fuel ratio for three different models. b) Corresponding ACT development with increasing air-fuel ratio.

Parameter	**Value**	**Parameter**	**Value**
n_{engine}	8000 rpm	$T_{\text{o,i}}$	364.15 K
t_{int}	306.15 K	$T_{\text{w,i}}$	356.5 K
t_{amb}	300.15 K	m_{fuel}	91.25 mg/stroke
λ_{cmb}	0.9 - 1.4	α_{ign}	26.0 - 42 °CA

Table 4.2: Engine parameters during variation of air-fuel ratio at a constant fuel mass flow rate.

For lean combustion states, the situation is the same for the average gas temperature. However, as can be seen in Fig. 4.12 b), after the transition to rich combustion, the temperature is almost unchanged. One possible reason for this phenomenon can be the calculation of the amount of molecules after the chemical reaction. In this chapter, the simplified water-gas shift reaction is used. As previously mentioned in Section 4.2, the amount of gas is directly used to calculate the temperatures. In reality, there can be different reaction paths of the elementary reactions with their corresponding kinetic laws, especially in rich combustion. According to (Warnatz et al., 2006) and (Koch, 2002), another possible reason can be the change in the composition of exhaust species and their heat capacities. As can be seen in Fig. 4.13 b), some improvements in the modified average gas temperature can be achieved by introducing the evaporation effect with Eq. (4.15).

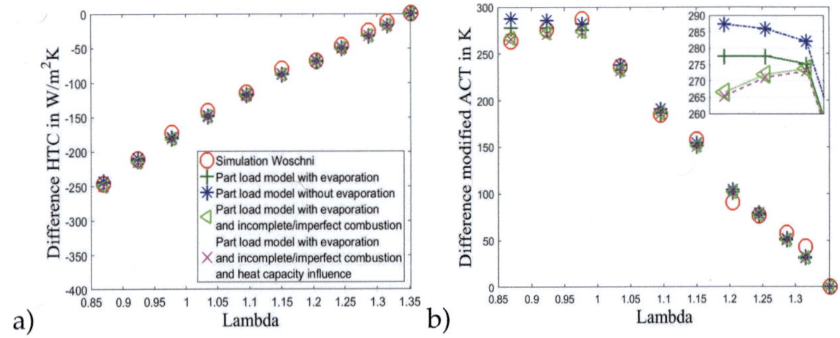

Figure 4.13: Progressions of the heat transfer coefficients and the modified average gas temperatures as functions of the air-fuel ratio. a) HTC increase in the combustion chamber for five different models: original Woschni, and the suggested part load model with successive activation of various effects: evaporation according to Eq. (4.15), incomplete combustion according to Eq. (4.14), and the influence of different heat capacities (Klein and Eriksson, 2004). b) Corresponding ACT development with increasing air-fuel ratio.

In lean combustion states, there is no much difference by including fuel evaporation. However, the modified average gas temperatures under rich combustion situations are closer to the original Woschni model. Similarly, the effect of incomplete and imperfect combustion, which is modeled according to Eq. (4.14), can better explain the change of sign of the gradient under rich combustion conditions. In consideration of differences in the heat capacity, no significant change can be reached. In this case, the ratio of heat capacities of the exhaust gas in the fourth stroke $\gamma_c(\overline{T}_g, \lambda_{\text{cmb}}) = c_p/c_v$ is modeled according to Klein and Eriksson (2004).

Afterwards, the heat capacity at constant volume is eliminated with the help of the universal gas constant. Nevertheless, the sensitivity is lower, compared to the original Woschni model. The increasing influence of radiation under rich conditions (Heywood, 1988), or the vague model Eq. (4.14), could be two additional factors. Moreover, changes in the volumetric efficiency by the boost pressure variation influence the effective gas mass in the cylinder. Deviations of fuel properties in Eq. (4.15) can also be a possible reason. The differences in the heat transfer coefficients are negligible small. In Fig. 4.14 a), simulated temperatures for the exhaust valve seat (seat ring) are compared with measured values. For air-fuel variations, the simulation is in good agreement with experimental data. The differences are in the range of measurement uncertainty.

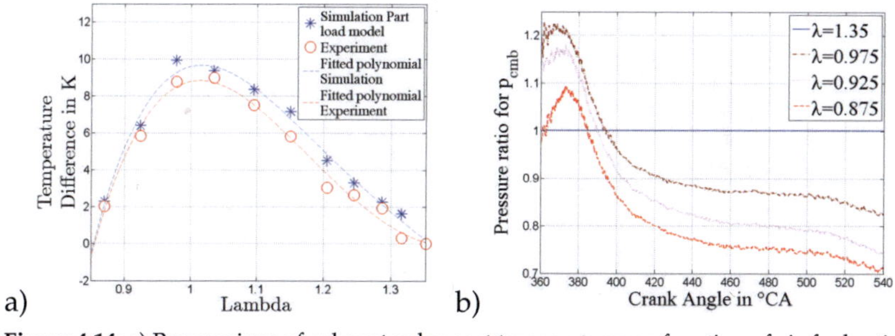

Figure 4.14: a) Progressions of exhaust valve seat temperatures as function of air-fuel ratio. The dashed lines are fitted polynomials of the fourth order using the method of least squares. b) Pressure due to combustion p_{cmb}. Values are instantaneously normalized to the pressure curve of highest lambda variation.

Corresponding ratios for the combustion pressure p_{cmb} are shown in Fig. 4.14 b). A noteworthy trend only occurs in rich combustion. In contrast to ignition time variations, after an initial overshoot, a more homogenous ratio can be observed.

4.3.3 Variation of boost pressure

In Fig. 4.15, the differences in the heat transfer coefficient and the modified average gas temperature for different loads are shown. The engine operation state is shown in Table 4.3.

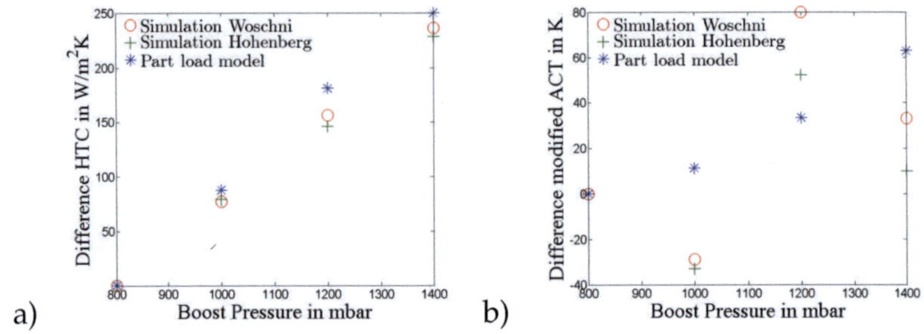

Figure 4.15: Progressions of the heat transfer coefficients and the modified average gas temperatures as function of boost pressure. a) HTC increase in the combustion chamber due to higher boost pressure for three different models. b) Corresponding ACT development with increasing boost pressure.

Parameter	Value	Parameter	Value
n_{engine}	8000 rpm	$T_{\text{o,i}}$	387.15 K
t_{int}	301.15 K	$T_{\text{w,i}}$	386.5 K
t_{amb}	294.65 K	m_{fuel}	27 - 52 mg/stroke
λ_{cmb}	1.21	α_{ign}	44.0 - 37.5 °CA

Table 4.3: Engine parameters during variation of boost pressure at a constant air-fuel ratio.

The highest boost pressure is chosen as a reference point for the part load model. Concerning heat transfer coefficients, there is a good agreement with the Woschni and Hohenberg model. Compared with the other two models, Fig. 4.15 a) shows only slightly higher values for the proposed part load model. However, the average gas temperature does not show the curved line which can be observed with the two detailed models. In the case of the part load model, it results in a continually growing gas temperature. In Fig. 4.16 a), simulated and measured cylinder head temperatures are shown. The difference has a maximum value of 1.75 K for a boost pressure of 1000 mbar. Fig. 4.16 b) shows corresponding ratios for the combustion pressure p_{cmb}. The assumption of a homogenous pressure ratio over crank angle is, therefore, nearly satisfied.

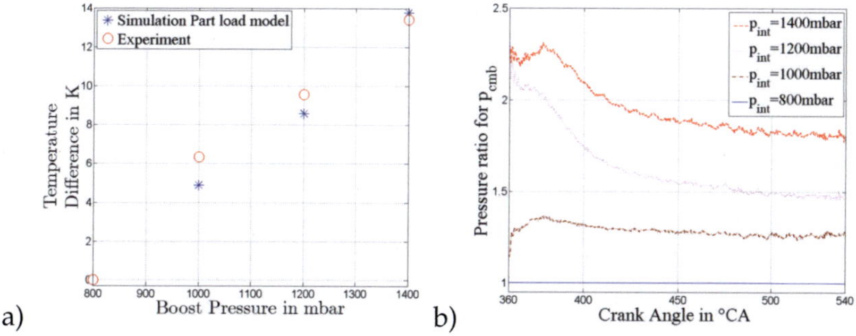

Figure 4.16: a) Progressions of cylinder head temperatures as function of boost pressure. b) Pressure due to combustion p_{cmb}. Values are instantaneously normalized to the pressure curve of lowest boost pressure variation.

A current research topic is the stratified, lean burn combustion technology with a drastically uneven equivalence ratio distribution. According to the measurements of Huegel et al. (2015), in relation to homogenous engine conditions, the Woschni model is able to predict the resulting differences in the heat transfer. Therefore, according to Eq. (4.9), the proposed method should also be suitable for stratified engine conditions. Validation tests remain to be carried out. Normally, according to Heywood (1988), various swirl and tumble ratios can be calibrated with the Woschni constant and the adjustment of the model parameters c_1 and c_2. The suggested method is independent from such modifications.

4.4 Conclusions

Concerning the initial research question, if a fast determination of thermal boundary conditions including only cycle integrated quantities instead of crank angle resolved in-cylinder pressure data is possible, the result is as follows.
With regard to changes in cycle averaged component temperatures, caused by variations in ignition time, air-fuel ratio and boost pressure, the proposed statistical model predicts sensitivities of heat transfer coefficients and reference temperatures which are comparable with common heat transfer models according to Woschni or Hohenberg. For lambda and load variations, simulated solid temperatures, which are based on the proposed thermal boundary conditions, are in good agreement with experimental data: temperature differences are in the range of measuring uncertainty. With regard to the turnaround from lean to rich combustion, the consideration of incomplete and imperfect combustion, as well as fuel evaporation should be taken into account. However, one can clearly recognize the model deficits in cases with a highly non-uniform pressure change over crank angle. From a practical point of view, ignition time variations only have significant pressure changes in the first third of the expansion stroke. Due to the fact that the proposed model only uses cycle integrated quantities instead of time-resolved data, such unbalances cannot be considered. Therefore, strong non-uniformities can lead to some errors in the heat flux and the resulting solid temperatures, as shown in the example of ignition time variations: the calculated temperature increase at the cylinder head, due to an advanced ignition time of 2.5 °CA, is underestimated by 2.5 K.

Chapter 5

Transient temperature calculation method for complex fluid-solid heat transfer problems with scattering boundary conditions[1]

5.1 Outline of the chapter

Regarding daily development work and the current available computing power, it is not possible to investigate every possible thermodynamic state of an engine during a transient drive with detailed 3D-CFD simulations, containing heat transfer treatments. Therefore, the research question can be formulated as follows: With regard to the dynamic response of engine component temperatures and the heat flux of the water jacket, is it possible to develop a transient calculation method which is a workable solution technique in the industrial practice? In particular, with respect to the different thermal behavior of various components and to variations in the engine setting like air-fuel ratios or water temperatures, how should the method be established in order to resolve the aforementioned frequency range of 0.1-1 Hz?

[1] This chapter is based on the paper "Transient temperature calculation method for complex fluid-solid heat transfer problems with scattering boundary conditions" (Hölz et al., 2019b) **DOI: https://doi.org/10.1016/j.applthermaleng.2018.12.081**

Using dimensional analysis, introduced in Chapter 2, and a statistical description of all relevant quantities, tested in Chapter 3, a MATLAB© - StarCCM+© interface is developed and implemented, which provides transient thermal boundary conditions for a subsequent three-dimensional finite volume simulation. By using in-cylinder pressure measurements under stationary conditions as an input, the method can account for different engine mappings or cylinder individual effects. Small deviations from these stationary states under a transient drive, resulting from diverse variations in ignition time or air-fuel ratio, are modeled with the aforementioned fast algorithm, which is described in Chapter 4. Due to restricted calculation time, a detailed CFD simulation of the water jacket, including turbulence modeling according to Chapter 2, cannot be performed under transient conditions. To calculate water heat transfer, a new method for determining heat transfer coefficients is, therefore, presented.
The method is validated with a four-cylinder, turbocharged SI engine. Therefore, thermocouples are installed near the intake and exhaust ports as well as the combustion chambers, including cylinder head and liner. Piston temperatures are evaluated by hardness measurements. Cylinder individual heat transfer phenomena, resulting from different volumetric efficiencies and unequal residual gas, are shown. Moreover, different transient behaviors of various measuring points around the combustion chamber are presented. Finally, a variation of the water inlet temperature and the resulting effect on solid temperatures is shown.

5.2 Method used in this chapter

The outer boundary conditions according to Eq. (3.1) will now be defined as a time-dependent, five-dimensional engine state matrix. Either, these quantities can be measured, or they can be modeled within the ECU. In the context of a three-dimensional, transient finite-volume simulation, one has to determine the inner boundary conditions; more specifically, the thermal boundary conditions. In the following, a method is suggested to translate *outer, concentrated* boundary conditions into *inner, distributed* boundary conditions.

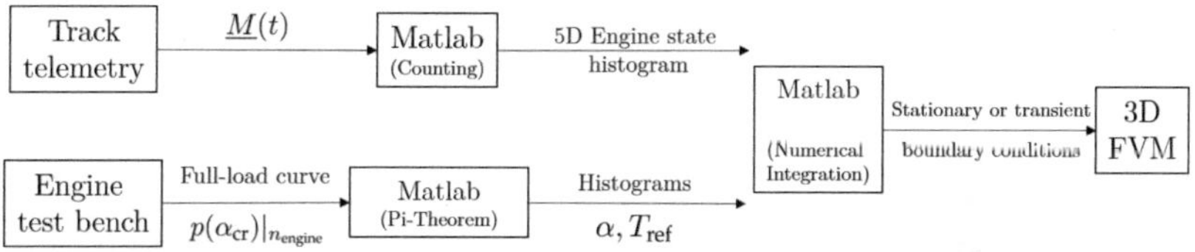

Figure 5.1: Method overview to generate boundary conditions for a 3D FVM simulation. Track data $\underline{M}(t)$ and pressure indication measurements $p(\alpha_{cr})$, as function of crank angle α_{cr}, are required. Usually, the reference temperature T_{ref}, which is related to the heat transfer coefficient α, is the in-cylinder gas temperture.

An overview of the method is shown in Fig. 5.1. With the help of a self-programmed code in MATLAB©, boundary conditions are determined for a subsequent 3D-FVM simulation of the full engine. Therefore, the engine state $\underline{M}(t)$ is imported by track telemetry with a high enough sampling frequency. All measured states are discretised in a 5D engine state matrix for which a histogram is computed. This histogram is normed in order to use it as an approximation for the probability density function $p_{\underline{n}}$ for the random variable $\underline{n}$ which describes the 5D engine state. Thus, in this case, the engine state is interpreted as a random variable. For stationary simulations, this density function can be used to get a kind of expectation field of the solid temperature $\langle T \rangle(\boldsymbol{x})$. This field can be used for comparison purposes between different engine mappings or race tracks. It also serves as an initial solution for subsequent transient simulations. In addition, one needs a high-pressure indication measurement $p(\alpha_{cr})$ of the in-cylinder gas pressure as a function of the crank angle α_{cr}. In this chapter, these stationary measurements are resolved with 0.1 °CA. The full-load curve consists of 14 engine speed points for which 60 complete engine cycles are recorded in order to take account of the scattering nature of combustion processes. Again, according to Eq. (3.2), thermal boundary conditions are assumed to be of the Newton form. Different models can be used to calculate the heat transfer coefficient with the help of the measured pressure data $p(\alpha_{\mathrm{cr}})$. Again, all quantities are interpreted as random variables. However, in this case one has to use conditional probability density functions. In the case of heat transfer coefficients, this is $p_{\alpha|\underline{n}}$ with $\underline{n}$ as the random variable of the engine state. One can easily show that the following relation holds for conditional probability density functions

$$p_{\alpha|\underline{n}}(A|\underline{N}) = p_{\alpha\underline{n}}(A,\underline{N})/p_{\underline{n}}(\underline{N}). \tag{5.1}$$

The realisations of the both random variables α and $\underline{n}$ are described with A and $\underline{N}$. The joint probability density function is given by $p_{\alpha\underline{n}}(A,\underline{N})$. It must be noted that the two random variables are strongly statistically dependent:

$$p_{\alpha\underline{n}}(A,\underline{N}) \neq p_{\alpha}(A)p_{\underline{n}}(\underline{N}). \tag{5.2}$$

Physically speaking, a heat transfer coefficient α strongly depends on the engine state $\underline{M}(t)$. The description as a random variable is quite useful, because one can pay attention to the cyclic fluctuations in the pressure curves as well as the intermittent operation of an engine, without simulating each stroke in detail. Therefore, this method implies a low-pass filter function which only resolves the lower frequencies in the range of the characteristic frequencies of the engine state $\underline{M}(t)$. High frequencies of the individual engine cycles are filtered out. Therefore, in the case of a four-stroke engine, a time discretization in the order of $2\ (1/n_{\mathrm{engine}})\ 60 \geq 0.015$ seconds is used. In this thesis, a race engine is investigated. As can be seen in Fig.5.4, most of the time, such a race engine is in the full-load or in a coasting state. That is the reason why only the full-load curve is measured. However, the part load model, which is introduced in Chapter 4, can be used for modeling probability density functions for engine states which significantly differ from the full-load or coasting state. In summary, one gets for any arbitrary random variable $\tilde{f}(\alpha, \underline{n})$ the following expressions for its expectation values. In case of a quasi-stationary simulation in order to calculate a kind of expectation values $\langle T\rangle(\boldsymbol{x})$:

$$\begin{aligned} \langle \tilde{f}(\alpha, \underline{n}) \rangle &= \int\limits_{\mathbb{R}^5_{\geq 0}} \cdots \int \int\limits_{\mathbb{R}_{\geq 0}} \tilde{f}(A, \underline{N}) p_{\alpha \underline{n}}(A, \underline{N}) \, dA \, d\underline{N} \\ &= \int\limits_{\mathbb{R}^5_{\geq 0}} \cdots \int \underbrace{\left(\int\limits_{\mathbb{R}_{\geq 0}} \tilde{f}(A, \underline{N}) p_{\alpha|\underline{n}}(A|\underline{N}) \, dA \right)}_{\langle \tilde{f}(\alpha, \underline{n}) | \underline{n} = \underline{N} \rangle} p_{\underline{n}}(\underline{N}) \, d\underline{N}. \end{aligned} \tag{5.3}$$

In case of a transient simulation in order to calculate time-dependent expectation values $\langle T\rangle(\boldsymbol{x}, t)$:

$$\langle \tilde{f}(\alpha) \rangle(t) = \int\limits_{\mathbb{R}_{\geq 0}} \tilde{f}(A) p_{\alpha|\underline{M}(t)}(A) \, dA. \tag{5.4}$$

Equation (5.3) and (5.4) apply to both, the wetted surfaces with gas, like valves, channels or the combustion chamber, as well as the water channel. However, as will be explained later in Section 5.4.2, the HTC for the water channel is assumed to be clearly determined for a given engine speed n_{engine}. In this case, the probability density function is the Dirac Delta-Distribution. It must be noted that $p_{\alpha|\underline{n}}(A|\underline{N})$ also contains coasting conditions, that means $T_{\text{i}} = 0$. The inner integral of Eq. (5.3) gives the conditional mean $\langle \tilde{f}(\alpha, \underline{n}) | \underline{n} = \underline{N} \rangle$ of the random variable $\tilde{f}(\alpha, \underline{n})$. Remember that, due to the non-linearity in Eq. (3.2), a statistically modified reference temperature $\langle \alpha T_{\text{ref}} \rangle / \langle \alpha \rangle$ is necessary. Whenever it will be spoken of a reference temperature, this modified version is meant.

In the following, the acronym ACT, for average cylinder temperature, is used. Remember that Eq. (5.3) and (5.4) also applies to the modified reference temperature.

5.3 Programming details

In Fig. 5.2, an overview of the code structure is shown. Beginning with the calculation of the probability density functions for the HTC and ACT with the help of pressures measurements, the analysis of the transient engine states follows.

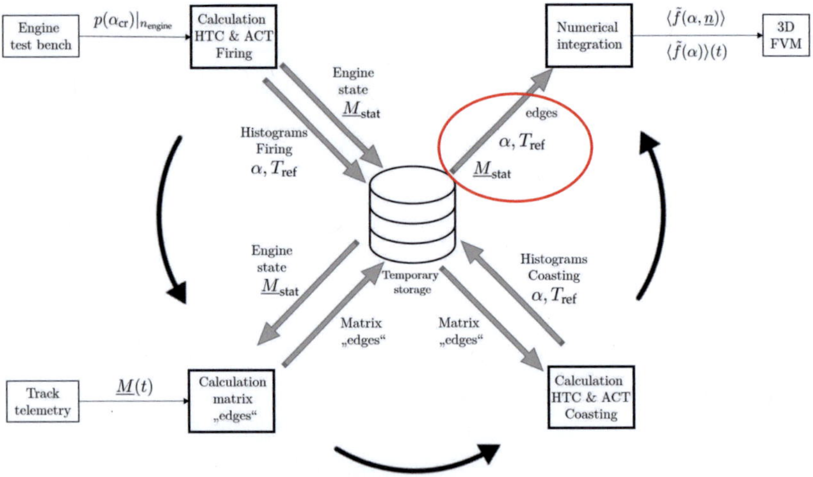

Figure 5.2: Programming overview.

The stationary matrix $\underline{M}_{\text{stat}}$ corresponds to the engine state matrix $\underline{M}(t)$ from Eq. (3.1) for the stationary measuring points. It serves as a reference state to which the transient states $\underline{M}(t)$ are related. Especially, using the part load model from Chapter 4, ratios of the matrix entries are of interest. Afterwards, for each discretized engine state, the coasting condition is modeled and the corresponding probability density functions are calculated. In the end, the PDF's are integrated in order to get quasi-stationary, $\langle \tilde{f}(\alpha, \underline{n}) \rangle$, or transient, $\langle \tilde{f}(\alpha) \rangle(t)$, boundary conditions for the subsequent FVM simulation. The probability density functions are implemented as normed histograms. Therefore, a discretization for each state variable, e.g., the inner and outer boundary conditions, is needed. Just like the modeled HTC and ACT, the engine state matrix $\underline{M}(t)$ is represented as a real matrix which is called *edges*. It consists of five dimensions: each of them consists of a vector with different number of elements, depending on the discretization level of the corresponding variable. It is important to note that the histograms have to be normed regarding the corresponding integration field. Multidimensional integration uses Fubini's theorem resulting in a piecewise integration over all dimensions of $\underline{M}(t)$. In case of Eq. (5.3), it follows:

$$\langle \tilde{f}(\alpha, \underline{n}) \rangle \approx \overbrace{\sum_{\underline{N}} \underbrace{\left(\sum_{A} \tilde{f}(A, \underline{N}) \hat{p}_{\alpha|\underline{n}}(A|\underline{N}) \, \Delta A \right)}_{f_{\text{slave}}(\underline{N})} \hat{p}_{\underline{n}}(\underline{N}) \, \Delta \underline{N}}^{f_{\text{master}}(\underline{N})}. \tag{5.5}$$

The corresponding normed histograms, with respect to both summations, are described by $\hat{p}_{\alpha|\underline{n}}(A|\underline{N})$ and $\hat{p}_{\underline{n}}(\underline{N})$. Note that the part load model only changes the function $f_{\text{slave}}(\underline{N})$ which has the same structure as the matrix *edges*. It is clear that lots of entries of $\hat{p}_{\underline{n}}(\underline{N})$ are zero, because some combinations are not physical: as an example, one cannot get a positive value for the induced torque $T_{\text{i}} > 0$ if no fuel is injected $m_{\text{fuel}} = 0$. Due to the state discretization, the function $f_{\text{slave}}(\underline{N})|n_{\text{engine}}$, evaluated for a specific engine speed n_{engine}, is received by linearization between corresponding test bench engine speed grid points $n_{\text{engine}}^{\text{left}}$ and $n_{\text{engine}}^{\text{right}}$:

$$f_{\text{slave}}(\underline{N})|n_{\text{engine}} = (1-a) f_{\text{slave}}(\underline{N})|n_{\text{engine}}^{\text{left}} + a f_{\text{slave}}(\underline{N})|n_{\text{engine}}^{\text{right}}, \tag{5.6}$$

with

$$n_{\text{engine}}^{\text{left}} \leq n_{\text{engine}} \leq n_{\text{engine}}^{\text{right}}, \tag{5.7}$$

and

$$a = \frac{n_{\text{engine}} - n_{\text{engine}}^{\text{left}}}{n_{\text{engine}}^{\text{right}} - n_{\text{engine}}^{\text{left}}}. \tag{5.8}$$

To perform transient simulations with Eq. (5.5), one needs the conditional mean $\langle \tilde{f}(\alpha, \underline{n})|\underline{n} = \underline{N}\rangle$. Of course, this value is a function of time, because the engine state matrix $\underline{M}(t)$ is a function of time. For this kind of simulation, a pointer matrix P is implemented which is a matrix with the same number of rows as points in time, which depends on the sampling rate of $\underline{M}(t)$ and the temporal discretization of the 3D FVM simulation, and five columns, corresponding to the five-dimensional engine state. These columns contain the position within the matrix $f_{\text{slave}}(\underline{N})$. The procedure is given in Fig. 5.3: for a given simulation time t_{Sim}, the corresponding matrix row n_{Sim} contains the pointer vector $P(n_{\text{Sim}}, :)$. It provides the necessary indices for the matrix f_{slave}. As an example, in Fig. 5.4 a), a projection of the five-dimensional engine state $\underline{M}(t)$ for a typical race lap is shown. The coasting state is described by the condition $T_{\text{i}} = 0$. Statistical analysis shows that about two third of the complete time the engine is in full load state, while about a quarter of the period the coasting condition is fulfilled. In Fig. 5.4 b) and c), the corresponding discretized sample space is shown. As one can see, in race conditions, the part load state is almost negligible, and, therefore, the model accuracy from Chapter 4 provides only limited advantages.

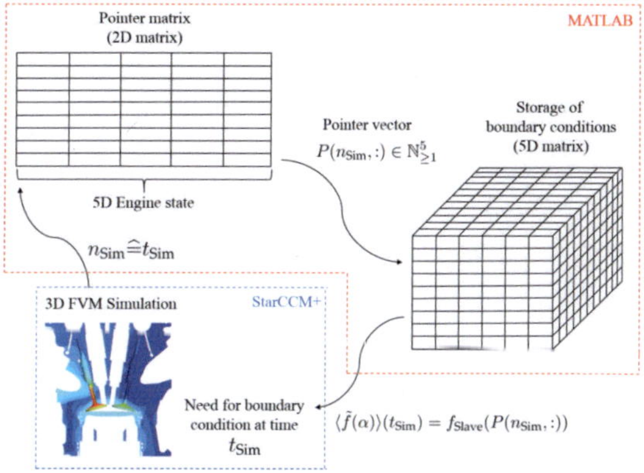

Figure 5.3: Transient simulation method. Boundary conditions are determined and saved in a MATLAB© environment. The subsequent finite volume simulation is done with the aid of the software StarCCM+.

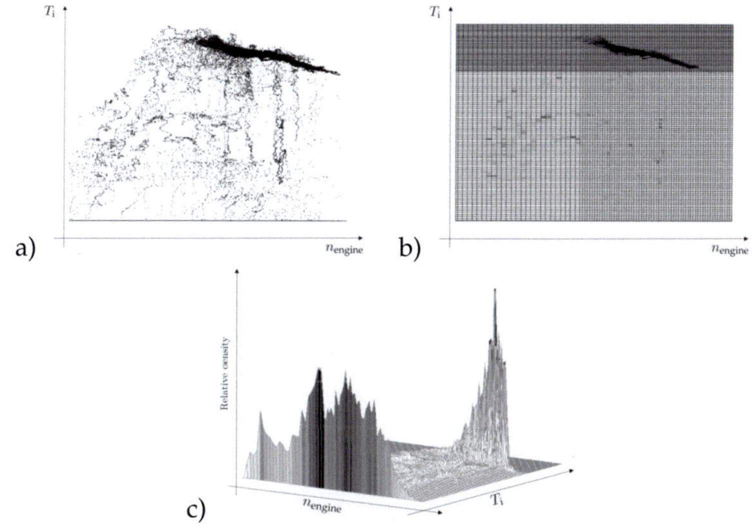

Figure 5.4: a) Exemplary scatter plot in the T_i-n_{engine} sample space for a typical race lap. b) Corresponding event counting in a discretized sample space with variable mesh size. c) Corresponding relative density of the events.

5.4 Determination of boundary conditions by means of similarity mechanics

For the piston, the thermal resistance model from Section 2.2.7 is used. The oil cooling boundary condition is set according to Eq. (2.34). However, due to experiences and validation data, it is found that the values for the HTC should be twice the proposed ones. Generally speaking, Table 2.3 delivers seven times larger values.

5.4.1 Transient heat transfer for gas-wetted boundaries

For the inlet and outlet system, the thermal boundary conditions from Section 3.3.1 are used. In case of the combustion chamber, the model according to Bargende is chosen for the heat transfer coefficient α (Bargende, 1991). For the lap-averaged results in Section 5.5.1, however, the model according to Woschni is selected (Woschni, 1967). Details to both models can be found in Section 2.2.5. Coasting conditions are determined with an isentropic assumption for the pressure during compression and expansion:

$$pV^{\kappa} = p_{\text{ini}} V_{\text{max}}^{\kappa}. \tag{5.9}$$

Therefore, using the first three entries of the five-dimensional engine state matrix $\underline{M}(t)$, the heat transfer coefficient under coasting conditions can be easily modeled. In this case, κ is the isentropic exponent. The subscripts denotes the initial pressure and the maximum volume. In literature, on can find various approaches for convective heat transfer equations under motored conditions. The influence of engine speed is investigated in Sanli et al. (2008). In Irimescu (2013) an alternative model for motored conditions is suggested and compared with conventional engine models. A review about this topic is given in Broekaert et al.

(2016). The effect of gas properties is explicitly investigated in Demuynck et al. (2012).

A full load state under race conditions differs from stationary measurements on a test bench. According to Fig. 5.4 a), it does not exist a line, but rather a small subspace of full load points. The ECU has to adjust permanently the operating conditions: ignition angle, fuel and air mass flow - to name just a few aspects. Moreover, inertia effects of several subsystems like turbochargers cause transient boundary conditions: generating the required boost pressure after coasting needs time resulting afterwards in some overshoots. Therefore, according to Chapter 4, the part load model can be used. An overview is given in Fig. 5.5. The probability density functions derived with measured pressure data from the stationary test bench serve as input. Depending on the transient engine state $\underline{M}(t)$, new probability density functions are modeled. The matrix entries of $\underline{M}(t)$, describing the transient engine condition according to Eq. (3.1), and $\underline{M}_{\text{stat}}$, the corresponding full load state at stationary conditions, are related. Let α_{stat} and $\alpha(t)$ be two realisations for the heat transfer coefficient in the state $\underline{M}_{\text{stat}}$ and $\underline{M}(t)$, respectively.

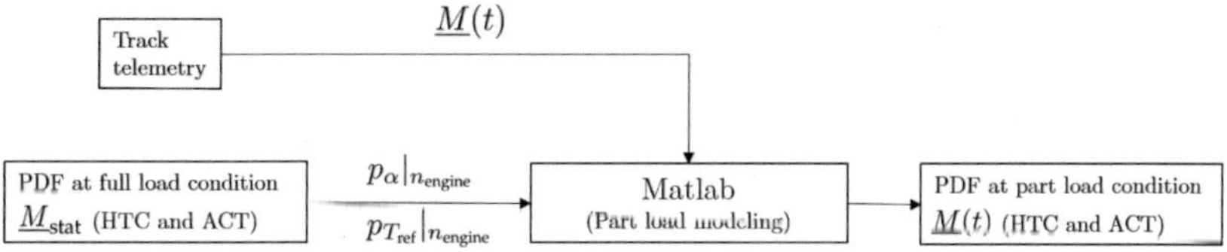

Figure 5.5: Method overview for generating probability density functions for the HTC and ACT at part load conditions. Given are the probability density functions at the full load condition and the engine state matrix $\underline{M}(t)$.

Using Eq. (4.7) for the transformation of realisations

$$\alpha(t) = \beta_T(\underline{M}_{\text{stat}}, \underline{M}(t), \alpha_{\text{stat}})\alpha_{\text{stat}}, \tag{5.10}$$

the new probability density functions according to Eq. (5.3) and (5.4) can be easily calculated. The transformation coefficient β_T can be described by Eq. (4.10)

$$\beta_T(\underline{M}_{\text{stat}}, \underline{M}(t), \alpha_{\text{stat}}) = \left(\frac{p|_{\underline{M}(t)}}{p|_{\underline{M}_{\text{stat}}}}\right)^m \left(\frac{v|_{\underline{M}(t)}}{v|_{\underline{M}_{\text{stat}}}}\right)^m \left(\frac{\overline{T}_g|_{\underline{M}(t)}}{\overline{T}_g|_{\underline{M}_{\text{stat}}}}\right)^{0.75-1.62m}. \tag{5.11}$$

In Fig. 5.6, an example for using the part load model is shown. The inlet temperature t_{int} is kept constant, resulting in a four-dimensional engine state matrix $\underline{M}(t)$ from Eq. (3.1).

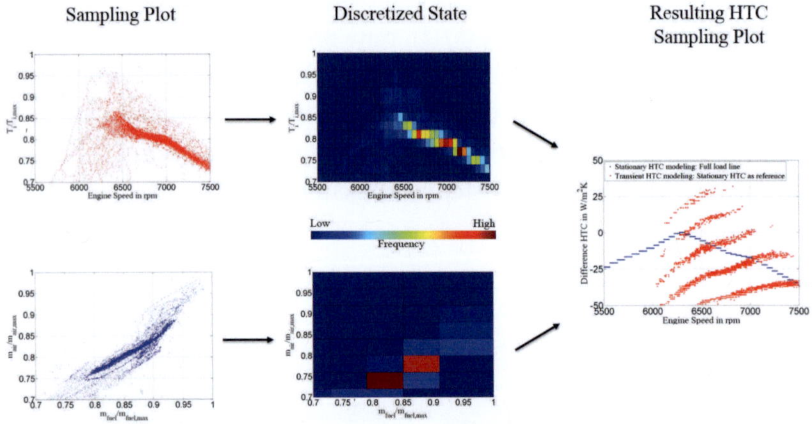

Figure 5.6: Example for using the part load model for a representative race lap. The T_{i}-n_{engine} and m_{air}-m_{fuel} sample spaces are shown and discretized with different mesh sizes. Values are normed to corresponding maximum values. The resulting HTC sample space is shown on the right side: values give the difference to the maximum, stationary HTC.

To visualize the different impacts, the m_{air}-m_{fuel} sample space is meshed with a much coarser mesh size than the T_{i}-n_{engine} sample space. The red separate subdivisions of the heat transfer coefficients result from the coarse mesh. The variation within one subdivision is the result of the T_{i}-n_{engine} sample space, which has a finer mesh. For comparison purposes, the full load line, resulting from stationary measurements on the engine test bench, is plotted as a blue reference line. The step function results from the engine speed discretization.

5.4.2 Transient heat transfer due to variations of water mass flow and temperature

According to similarity mechanics, one can assume a correlation of the form $Nu = f(Re, Pr)$ for the water channel. Because of the large heat capacity of water, the temperature in the water jacket varies quite slowly. Its difference during one lap is in the range of 5 K. Therefore, the dependence on the Prandtl number Pr can be neglected. With the same argument, the viscosity in the Reynolds number can be ignored, resulting in a pure dependence of the water velocity. Actually, it is very expansive to simulate nearly every occurring water mass flow rate during one lap within a detailed CFD simulation. In addition, if one wants to investigate solid temperature curves with transient simulations one ought to map the heat transfer coefficients at every time step on the solid mesh, resulting in an extremely high effort. Alternatively, one could simulate the fluid region with the transient Navier-Stokes equations simultaneously. However, due to turbulent flow with thin boundary layers, much more cells would be necessary in comparison with a pure solid simulation. Therefore, a new method is presented by simulating heat transfer by means of CFD for only one reference water mass flow rate. The result of this stationary reference simulation is a heat flux vector for each solid cell, e.g., a field function $\boldsymbol{q}_{\mathrm{ref}}(\boldsymbol{x})$. Assuming a correlation $Nu = f(Re, Pr)$ with a Reynolds exponent m, following separation approach for the HTC in transient simulations can be formulated:

$$\alpha(\boldsymbol{x},t) \approx \tilde{c}_{\text{ref}}(\boldsymbol{x}) v_{\text{water}}^{m}(t) \sim c_{\text{ref}}(\boldsymbol{x}) n_{\text{engine}}^{m}(t). \tag{5.12}$$

The proportionality factors $\tilde{c}_{\text{ref}}$ and c_{ref} contain implicitly the field function $\boldsymbol{q}_{\text{ref}}(\boldsymbol{x})$ from the stationary reference simulation. v_{water} is a characteristic velocity of the water channel flow. Due to the water pump and the fixed transmission ratio with respect to the engine, this velocity is proportional to the engine speed. Consequently, the transient HTC can be approximated with

$$\alpha(\boldsymbol{x},t) \approx \alpha_{\text{ref}}(\boldsymbol{x}) \left(\frac{n_{\text{engine}}(t)}{n_{\text{engine}}|_{\text{ref}}} \right)^{m}. \tag{5.13}$$

$n_{\text{engine}}|_{\text{ref}}$ is the engine speed in the stationary reference simulation. This reference engine speed should be chosen carefully according to the weighted average

$$n_{\text{engine}}|_{\text{ref}} = \left(\langle n_{\text{engine}}^{m} \rangle \right)^{1/m}. \tag{5.14}$$

Due to the isothermal assumption in Eq. (5.13), temperature dependencies of the form $\propto \lambda/\nu^{m} Pr^{n}$ are neglected. In this case, the Prandtl exponent is described with n. The mapped reference heat transfer coefficient α_{ref} is determined by

$$\alpha_{\text{ref}}(\boldsymbol{x}) = \frac{\boldsymbol{q}_{\text{cond}}(\boldsymbol{x}) \cdot \boldsymbol{n}(\boldsymbol{x})}{(T_{\text{ref}} - T_{s}(\boldsymbol{x}))}. \tag{5.15}$$

In this case, $\boldsymbol{q}_{\text{cond}}$ is the heat flux vector due to heat conduction in the solid part, and $\boldsymbol{n}$ is the boundary normal vector. Moreover, T_{ref} is the specified reference temperature, e.g., the water inlet temperature, and T_{s} is the solid temperature of the wall next cell. The determination of the reference HTC $\alpha_{\text{ref}}(\boldsymbol{x})$ is based on a CFD-CHT calculation method described in Section 3.4.

5.5 Results and discussion

According to the heat transfer model according to Bargende and Woschni, the necessary model calibration is done with a single cylinder engine. Therefore, one representative, stationary measuring point on the full load line is used: the engine speed is $n_{\text{engine}} = 6000$ rpm. Afterwards, no modifications of the model parameters are carried out. For model validation, transient measurements on a full engine are used. Both engines are geometrically equal with respect to relevant sections like the water channel, the combustion chamber or the gas channels. 70 measuring points are mounted on a four-cylinder, turbocharged engine, which is based on a Porsche racing application.

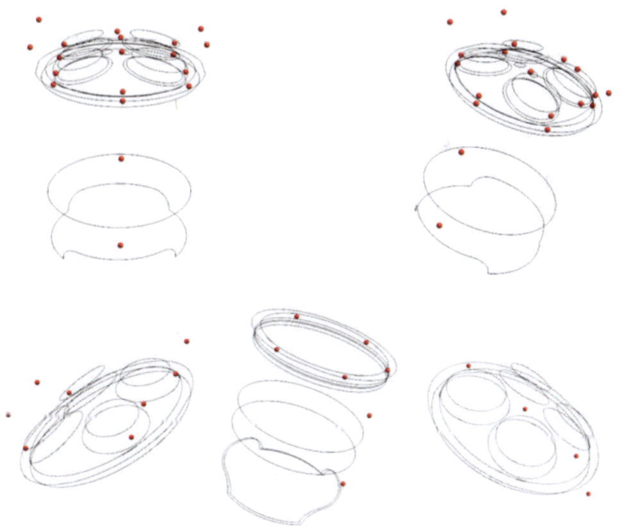

Figure 5.7: Exemplary measuring points for one cylinder in red. For a clear presentation, some component edges are plotted. Upper row: Overview of all measuring points. Lower row left: Inlet and outlet channels as well as rings. Lower row centred: Liner shoulder with axial variation. Lower row right: Combustion chamber wall of cylinder head with spark plug and intermediate measuring point in between two cylinders.

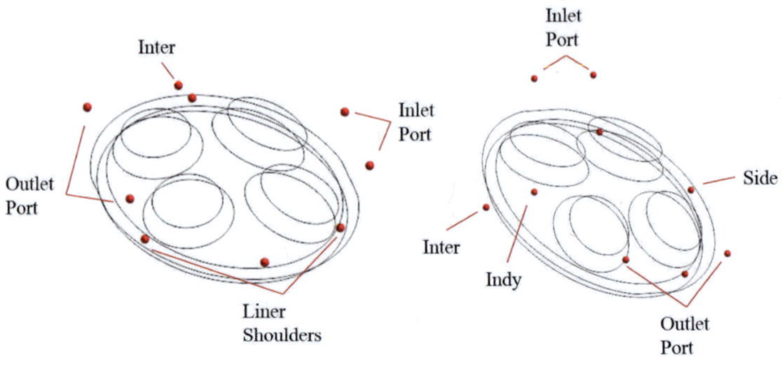

Figure 5.8: Temperature measurement positions at the cylinder head and the cylinder liner shoulder.

For solid temperatures, type K thermocouples with an uncertainty of 1 K are used. A heat-conducting paste with a thermal conductivity of 2.5 W/(mK) is put between the solid surface and the thermocouples. Fluid temperatures are measured with Pt100 sensors, which have an uncertainty of 0.053 K. Water temperatures are measured at the inlet and outlet. According to Eq. (5.15), these quantities are used as reference temperature for each time step during the transient simulation. In addition, a measuring turbine for the volume flow rate is installed. In Fig. 5.7 and 5.8, some measuring points for one cylinder are shown. The holes for all measuring points ended one millimeter below the surface.

5.5.1 Averaged temperatures

In Fig. 5.9 and 5.10, some lap-averaged temperatures are shown. The conditions are summarized in Table 5.1.

Parameter	Value	Parameter	Value
λ_{cmb}	1.095	max. m_{fuel}	80.6 kg/h
t_{int}	293.15 K	$T_{\text{w,i}}$	373.65 K

Table 5.1: Test parameters for validation purposes.

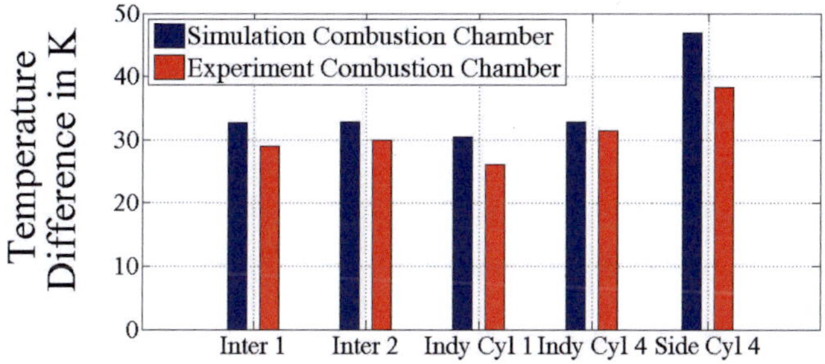

Figure 5.9: Temperature values near the combustion chamber. The temperatures are given in relation to the water inlet temperature. Positions: See Fig. 5.8.

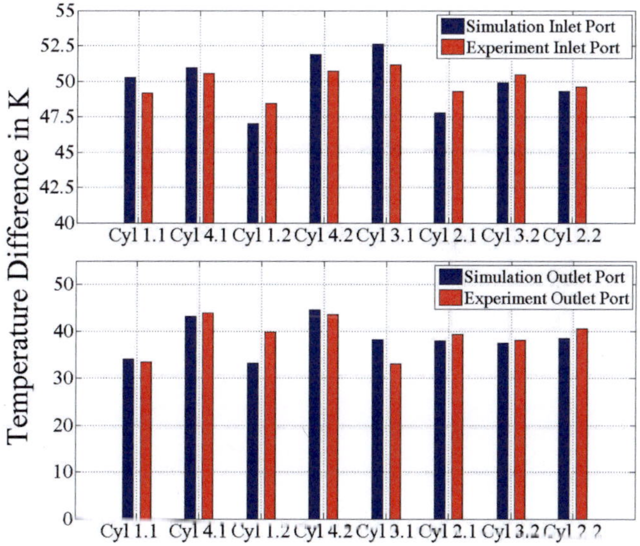

Figure 5.10: Temperature values for the inlet and outlet port. The exhaust port temperatures are given in relation to the water inlet temperature. The intake port ones in relation to half of the water inlet temperature. Positions: See Fig. 5.8.

Given are five different positions near the combustion chambers, which are allocated to four cylinders. The naming convention is explained in Fig.5.8. The names "Inter 1" and "Inter 2" refer to positions between cylinder one-two and three-four. Fig. 5.10 shows the results for the intake and exhaust ports. In general, the simulation is in good agreement with experimental results. In addition, the absolute level of the temperature can be well represented: remember that a single cylinder engine, without multiple positions, is used for calibration. However, by example of cylinder four, the temperature measurement at the side position is generally lower than the simulated value. These positions are located more far away from the water jacket, and are positioned near the outer surface of the engine. One possible reason is the underestimation of the heat transfer resulting from free convection. In this thesis, a constant value of 30 W/(m^{2}K) is used. Alternatively, the more linear assumption of Fig. 2.4 could be too conservative.

According to Fig. 5.11, the piston temperatures are evaluated with the help of hardness measurements. This method is state of the art. It is based on material aging curves which are determined experimentally. The underlying thermally activated processes can be described with Arrhenius relationships.

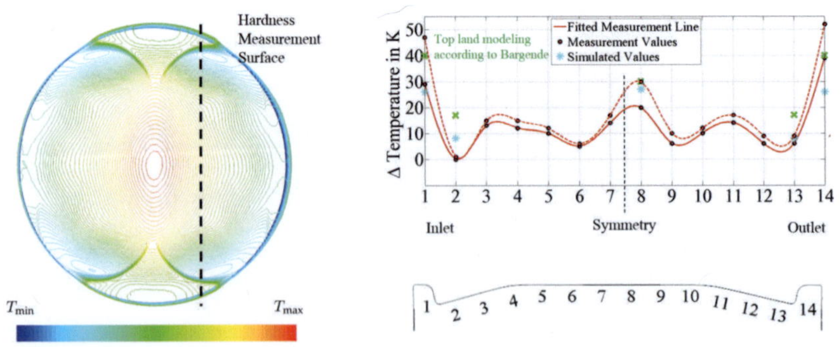

Figure 5.11: Relative temperature field of a piston crown and comparison with measurement results. Position 1 to 14 are used for hardness measurements. All values are referred to the minimum temperature on the measurement surface. The dashed line corresponds to a pure full load evaluation. The solid line results from a complete life cycle analysis, including the temporarily coasting states.

However, due to some uncertainties concerning hardness measurement techniques, the exact measurement position, and the scattering nature of material aging parameters, this validation method contains some errors: the absolute temperature vary approximately between $\Delta T = \pm 10$ K. The simulation predicts the maximum temperature in the middle of the piston with oval shaped isolines. In this case, a more linear assumption according to Fig. 2.5 is used. Therefore, the slight asymmetry towards the hotter exhaust side cannot be simulated. Similar results are reported in Esfahanian et al. (2006) or Choi et al. (1997). In addition, as can be seen in Fig. 5.11, the pronounced temperature increase on the piston sides cannot be simulated with such simplified spatial distributions. However, as can be seen from the green dots, an improvement can be made by using the piston top land modeling according to Bargende et al. (2011). In this work, at the other extreme, a thermodynamic equilibrium between the gas temperature and the solid temperatures of the surrounding surfaces is assumed. It is, however, to be considered that the global temperature minima also increase. Nevertheless, because of increased measurement uncertainties due to large temperature gradients, expecially at the exhaust side, this top land modeling seems to be expedient. For all subsequent investigation, especially the fatigue evaluation in Chapter 7, this model is used. Moreover, the local minima at position 6 and 9 cannot be rediscovered in the simulation. One theory for these local minima are inhomogeneously distributed heat fluxes on the bottom surface, resulting from complex oil flows and the direct oil injection. All in all, the open issue still remains the oil cooling: the discrepancies between the various empirical correlations in Section 2.2.7 are tremendous. Model input parameters differ significantly from each other, according to the dominant flow structure, and besides, individual piston geometries cannot be modeled. Therefore, the heat transfer coefficient remains, to some extent, a calibration parameter. Fig. 5.12 shows the effect of oil cooling: without an oil supply, there is a kind of temperature localization at the centre of the piston.

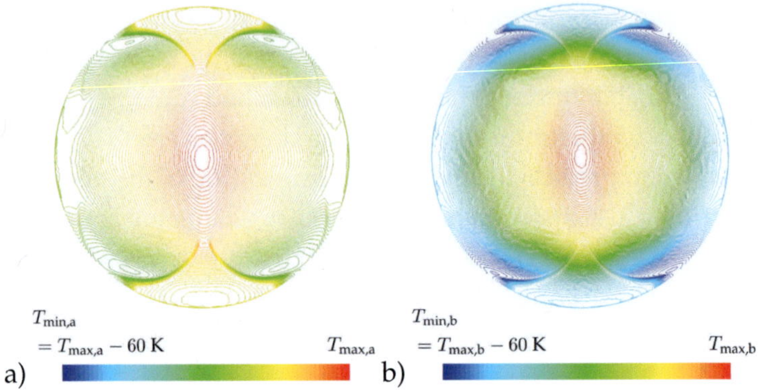

Figure 5.12: Effect of oil cooling on the piston crown.

The temperature gradients with oil cooling, Fig. 5.12 a), are only half the differences without oil cooling, Fig. 5.12 b). The maximum piston temperature $T_{max,a}$ with oil cooling is about 100 K lower than $T_{max,b}$, the maximum piston temperature without oil cooling.

5.5.2 Transient temperatures and heat fluxes

As a further example, a transient drive with the test parameters presented in Table 5.1 is investigated. Firstly, the water heat transfer is shown in Fig. 5.13. Due to some uncertainty of experimental measurements, the confidence interval is given. It is calculated by the extremes of the temperature differences. These are measured by a PT100 sensor which have an uncertainty of 0.053 K. As one can see, the maximum simulated heat flow is within the range of measurement inaccuracy. The heat flow is calculated with the volume flow rate, the heat capacity and the temperature difference between inlet and outlet. In the detail section below, one can see a slightly phase shift within the gear changes. The differences between simulation and experiment is about 10 % at the beginning. On the left side, raw data are plotted. Because of the thermal inertia effect of sensors, measurement quality can be improved by correcting water temperatures according to $T_{cor} = T + \tau_c \, dT/ \, dt$.

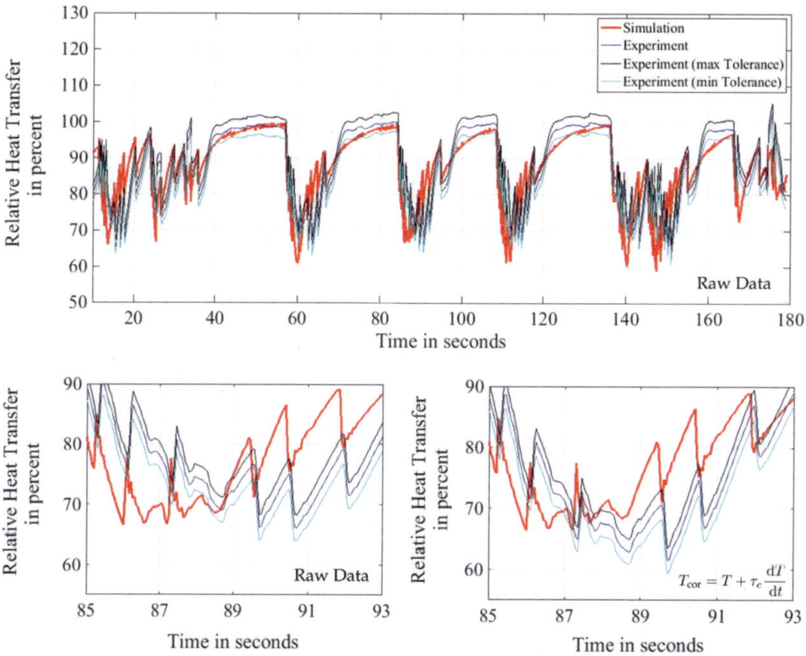

Figure 5.13: Above: Complete transient water heat transfer for one race lap. The settling time of about 10 seconds is not presented. Values are referred to the maximum measured heat flow. Below left: Detail section of the third acceleration phase including some gear changes. Raw data are shown. Below right: Same detail section. Measured temperatures are modified according to $T_{\text{cor}} = T + \tau_c \, \mathrm{d}T/\,\mathrm{d}t$.

The time constant τ_c can be determined by measuring the step response time. However, heat flow measurements have some potential sources of errors: Besides the measurement uncertainty of volume flow rates, the high sensitivity to temperature changes is challenging. Because of the large heat capacity of water, a small temperature change results in a high difference in heat flow. In addition, the measuring positions of the sensors make some differences. In transient experiments, there is another aspect: strictly speaking, simultaneous measurement of inlet and outlet temperatures is not right. That is the reason for the slightly phase shift in Fig. 5.13: the water takes some time to flow through the

complete engine. Moreover, pulsating flow can increase or decrease heat transfer in turbulent pipe flows. An analytical study about this problem can be found in Moschandreou and Zamir (1997) for laminar flows. In addition to the Reynolds number, there are much more system parameters which influence heat flow: Prandtl number, oscillating frequency and amplitude as well as the entry length of pipes. Experimental results with comparable parameters prevailing in engine water channels can be found in Patel and Attal (2016) or Habib et al. (2004). Looking at the results $Pr \approx 1$, $Re \approx 10^4 - 10^5$ and the oscillating frequency in the range of $1 - 5$ Hz, the Nusselt number increase, respectively decrease, is about 10 %. This can explain some discrepancies during gear changes.
Secondly, some solid temperature curves are presented in Fig. 5.14. Interestingly, one can observe a different thermal behavior for each engine section. The temperature at the inlet channel is quite constant and the differences between simulation and measurement are negligible small. In the initial phase within the first 30 s, the simulation shows a settling process. For the outlet channel, one can observe a strong degressive character for the temperature curve within acceleration phases. The temperature amplitudes are in the order of 20 K. The simulation shows weak over- and undershoots of about 3 K. However, the mean value is exact enough. The inlet rings show a step-wise linear system response for different time periods. Again, mean values are similar for measurement and simulation. Temperature measurements at the outlet rings are very challenging: normally, the thermal conductivity is very low resulting in high temperature gradients within a ring. Typical values are about 50 K/mm. Small position deviations of the thermocouple result in high temperature discrepancies. In addition, it is a strong intrusive measurement method. The high thermal conductivity of thermocouples disturbs the temperature field and increases the heat flow from the outlet rings. Nevertheless, simulated mean values are comparable with measurements. However, the different transient behavior is noticeable.

The simulation shows a continually growing temperature, whereas the measured line is piecewise constant.

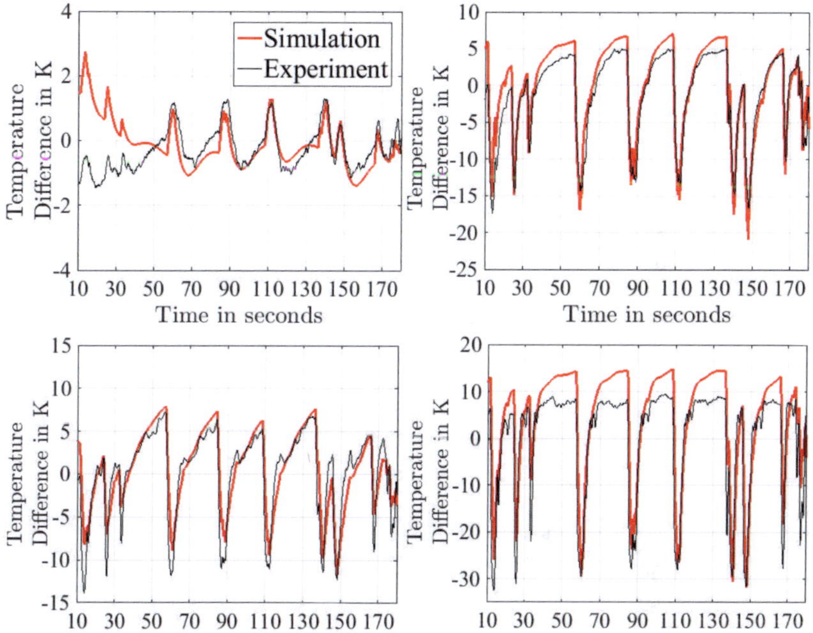

Figure 5.14: Simulated and measured temperature curves for various measurement positions. As reference temperature, the measured mean values are chosen. Above left: Inlet channel. Above right: Outlet channel. Below left: Inlet ring. Below right: Outlet ring.

5.5.3 Cylinder individual effects

Because of an unsymmetrical ignition sequence for the investigated engine, there are different volumetric efficiencies with unequal residual gas. In Fig. 5.15, two cylinder liner shoulders with different thermal behaviors are presented. For both cylinders, mean values as well as amplitudes of the simulated curves are in good agreement with experimental results. Over- or undershoots are temporarily in the range of

5 K. It is more interesting that the presented method is able to predict the characteristic curves for each cylinder. Observing an increasing temperature with a subsequent temperature drop at high-speed sections for the first cylinder, one can recognise a continual growth at cylinder four.

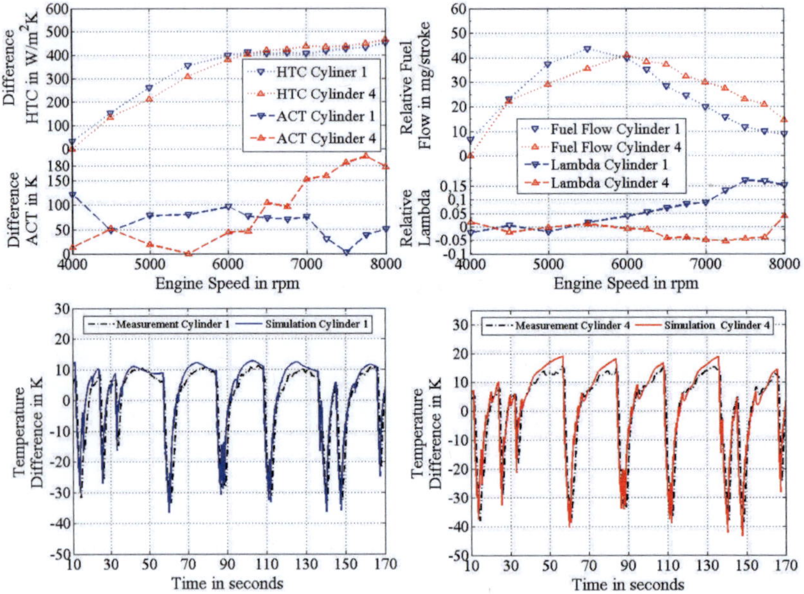

Figure 5.15: Different thermal loads for cylinder liner shoulders due to individual air-fuel ratios resulting from unequal residual gas. Above right: Injected fuel quantity and lambda value under stationary full load conditions. The lowest fuel flow for cylinder four as well as the mean lambda for both cylinders at 4000 rpm are used as a reference. Above left: Calculated HTC and ACT for both cylinders under full load state. In each case, the minimum values serve as a reference. Below: Simulated and measured transient thermal behavior for both cylinders. Temperature differences according to measured mean values are plotted.

As can be seen in Fig. 5.15, the main reason is the contrary progression of average cylinder gas temperatures with increasing engine speed. Cylinder four has a more stoichiometric combustion, whereas cylinder one gets higher air-fuel ratios resulting in lower gas temperatures. This is physically meaningful, because it is well known that the laminar flame speed and the adiabatic flame temperature have their maximum values when λ_{cmb} is a little bit lower than one (Warnatz et al., 2006). Remember that Fig. 5.15 only shows average gas temperatures. According to Eq. (3.3), the statistically modified temperature $\langle \alpha T_{ref} \rangle / \langle \alpha \rangle$ is needed. For quantitatively accurate simulations, the progression of the HTC is at least as important as gas temperatures. In this case, the qualitative progression is the same for both cylinders. Cylinder one has some slightly smaller values for high engine speed sections.

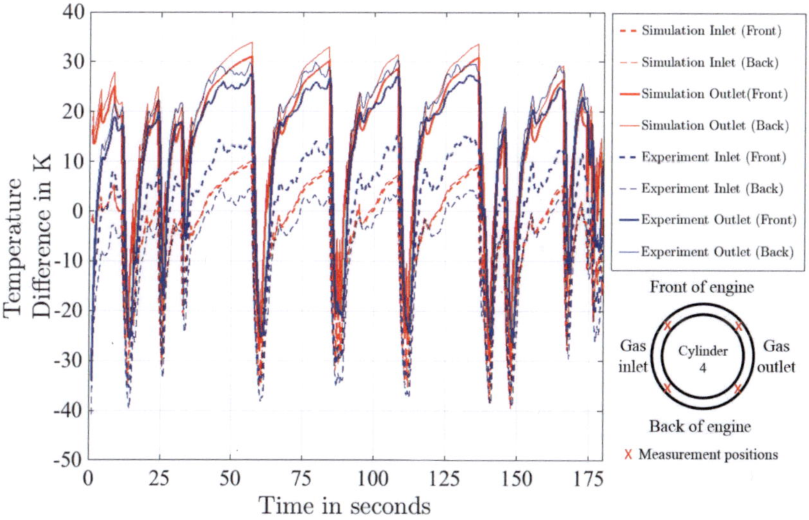

Figure 5.16: Temperature curves for different measuring positions at the cylinder liner shoulder of cylinder four. The mean temperature of the measured inlet temperature, at the front of the engine, is used as a reference.

In addition, Fig. 5.16 shows some spatial differences at the cylinder liner shoulder. Measuring positions near the exhaust side have about 20K higher temperatures at the end of straight. The main reason is the lower heat dissipation because of the hot exhaust channel of the cylinder head.

5.5.4 Variation of water quantities - simplified calculation method

Variation of water mass flow rate

To validate the proposed calculation method for transient water flows from Section 5.4.2, stationary measurements with different water mass flow rates are compared with simulation results. Therefore, the engine speed is kept constant. A summary of engine parameters is given in Table 5.2. As a reference water mass flow rate, a value of 2.32 kg/s is chosen. The water outlet Temperature is $T_{w,o}$.

Engine parameter	Value	Engine parameter	Value
n_{engine}	7000 rpm	$T_{o,i}$	364.15 K
t_{int}	310.15 K	$T_{w,o}$	361.65 K
t_{amb}	301.15 K	m_{fuel}	100 mg/stroke
λ_{cmb}	1.14	α_{ign}	29 °CA

Table 5.2: Engine parameters during the variation of the water mass flow rate

In Fig. 5.17, relative differences in the heat flux and the exhaust ring temperature are shown for two different model parameters m in Eq. (5.13). In literature, different values can be found: Taler and Taler (2017) investigated different forms for the Nusselt correlation in simple turbulent tube flows. They showed that the exponent strongly depends on the Prandtl number: increasing Prandtl number results in higher exponents. Water temperatures about 373 K are typical in engine applications. Using the proposed model in Taler and Taler (2017), values slightly larger than 0.8 should be expedient. In comparison with this model, experimental data by Eiamsa-Ard and Promvonge (2010) show only a little bit lower values for the Nusselt number gradient. However, both results are based on a tube length of 1250 mm. Uchida et al. (2005) showed that the Reynolds exponent m gets smaller with shorter tube length. This is in accordance with the results presented in this chapter. Using Reynolds number exponents in the range of 0.7, instead of 0.8, much better results can be achieved. Because of the complex geometry of engine water channels, models based on shorter tubes seems to be more expedient. Strictly speaking, the exponent also depends on the surface roughness and the Reynolds number itself.

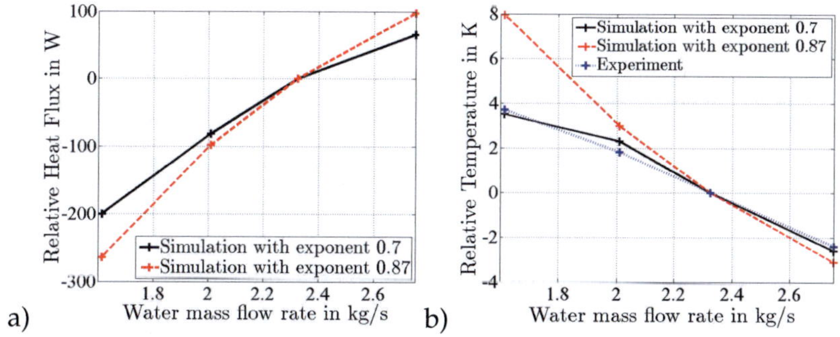

Figure 5.17: a) Relative heat flux to water concerning reference water mass flow rate at one cylinder. Two different Reynolds number exponents for Eq. (5.13) are shown. b) Resulting relative temperatures concerning reference water mass flow rate at the exhaust valve ring.

Higher values for the Reynolds number exponent result in a higher sensitivity to the relative heat flux concerning the reference water mass flow rate. In Fig. 5.17 a), the difference in the heat flux to the water channel is shown for two simulations with values $m = 0.7$ and $m = 0.87$. The resulting solid temperatures are shown in Fig. 5.17 b). As already mentioned, values higher than 0.8 overpredict the sensitivity of the Nusselt number. Using the proposed method from Section 5.4.2, a value in the range of 0.7 seems to be expedient. The corresponding simulated solid temperatures are in good agreement with the experimental results. In Fig. 5.18, a comparison of the field function $\alpha(\boldsymbol{x})$, according to Eq. (5.13), is shown.

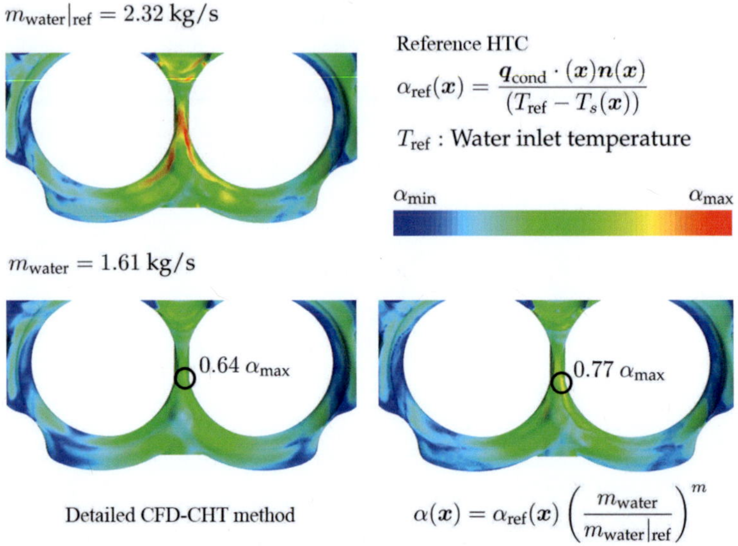

Figure 5.18: Water channel around the exhaust valve seats: comparison of the field function $\alpha(\boldsymbol{x})$ for a water mass flow rate of 1.61 kg/s. Above: Reference HTC for a water mass flow rate of 2.32 kg/s. Lower left: HTC field resulting from a detailed CFD-CHT method with $k - \omega$ turbulence model and wall laws for velocity and temperature. Lower right: HTC field resulting from the simplified method according to Eq. (5.13). The Reynolds exponent m is set to 0.7. Values give the relation to the maximum value of the reference simulation.

Quantitatively, the simplified method overpredicts the HTC values in the intermediate area of the valve seats by about 20 %.

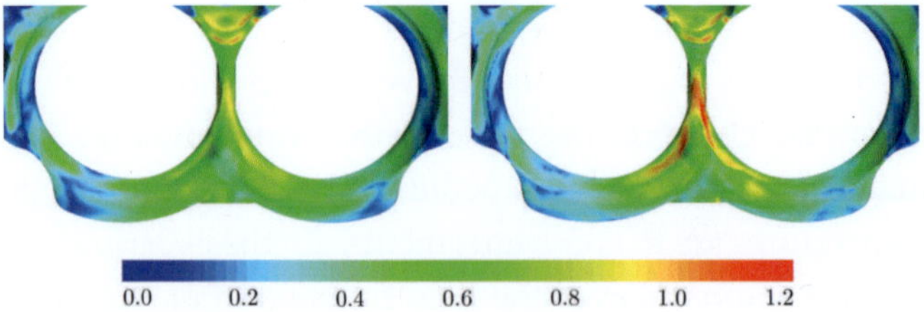

Figure 5.19: Same plot as Fig. 5.18. For a better comparison, values give the relation to the maximum value of the detailed CFD-CHT method for a water mass flow rate of 1.61 kg/s.

However, in some regions, e.g., several small areas on the side walls, the values are a bit lower than the detailed simulation.

Variation of water inlet temperature

Beside the HTC, the reference temperature in Eq. (3.2) should be noted. Therefore, transient measurements with four different water inlet temperatures $T_{w,i}$ are compared with corresponding transient simulations. Test parameters are given in Table 5.3.

Parameter	Value	Parameter	Value
λ_{cmb}	1.095	max. m_{fuel}	80.6 kg/h
t_{int}	296.15 K	$T_{w,i}$	various

Table 5.3: Test parameters during variation of water inlet temperature.

As an example, temperature curves for two different water inlet temperatures are shown in Fig. 5.20 a). As one can see, a change in water inlet temperature results in a kind of temperature offset.

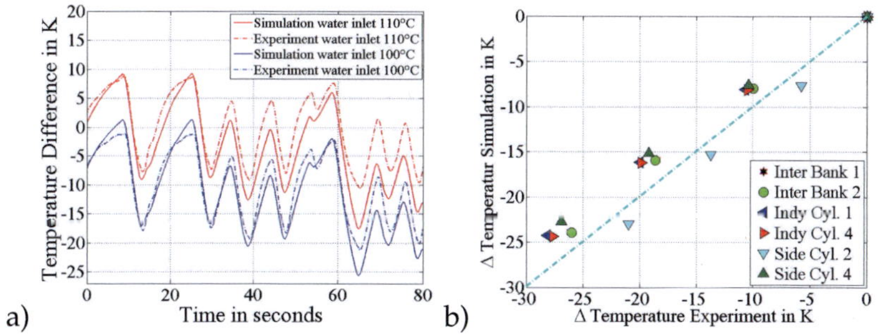

Figure 5.20: a) Simulated and measured temperature curves for the intermediate measuring point in between cylinder one and two. The mean value of the measured curve ($T_{w,i}$=383 K) serves as a reference for all four curves. b) Change of mean values for various measuring points at the combustion chamber wall: Inlet water temperature is progressively reduced by 10 K. The reference point is $T_{w,i}$=383 K. The dashed blue lines serve as an orientation and correspond to perfect simulations.

In case of Fig. 5.20 a), this offset is a little bit lower than the change of the water inlet temperature. Except of the first two acceleration phases, the simulated curves are below the measured lines. In Fig. 5.20 b), the change of temporal mean values for various measurement points of the cylinder head are shown. The distinct three collections of dots correspond to a reduction of the water inlet temperature from $T_{w,i}$=383 K to $T_{w,i}$=353 K. Measurement points at the combustion chamber wall of cylinder one, two and four are shown. A generally high sensitivity to water inlet temperature can be observed. However, most of the points are slightly above the dashed blue orientation line. For these points, the simulation underestimates the sensitivity. According to Fig. 5.8, the intermediate positions of the engine bank one and two correspond to points between two cylinders. The Indy points are radially positioned at approximately two thirds of the cylinder diameter. The corresponding Side points are around the circumference.

5.6 Conclusions

In this chapter, a transient calculation method for complex fluid-solid heat transfer problems is presented. Concerning the initial research question, how should a transient calculation method be established in order to simulate engine temperatures in the industrial practice, the result is as follows.

With the help of dimensional analysis and a stationary single cylinder engine calibration, the proposed method can predict solid temperatures accurate enough in the range of 0.1-1 Hz within a reasonable calculation time: using 80 CPUs, a complete full engine, which is meshed with 13.5 million cells for the solid parts, can be simulated within three hours. The corresponding simulation time is one race lap with 180 s. Temperature mean values and amplitudes are in good agreement with experimental data. Different thermal behaviors of various engine components,

cylinder-individual temperature effects, as well as the transient heat transfer within the water jacket can be simulated. Consequently, this work complements current simulation techniques in the field of transient thermal analysis of combustion engines in such a way that frequency ranges of 0.1-1 Hz can be successfully simulated.
Concerning the water jacket, at the end of straight, the difference between simulation and measurement for the heat flow is within the measurement tolerance. Under transient conditions during gear shifts, the error is less than 10 %. Differences between the detailed and the simplified method in the local heat transfer coefficient near the valves are found to be in a maximum range of 20 %. Overestimations, or underestimations respectively, of diverse temperature amplitudes are typically in the range of 2-4 K. Temporally, some values can be up to 7 K.

Chapter 6

Influence of various engine settings on cycle-averaged component temperatures under dynamic operating conditions

6.1 Outline of the chapter

The research questions can be formulated as follows: with regard to different engine settings, what is the effect on the dynamic temperature behavior of various engine components? In addition, following the method used in Chapter 3 for stationary, cycle-averaged solid temperatures, how purposeful are fast and simplified heat transfer models under dynamic operating conditions? In particular, depending on the impact of each specific, thermodynamic variation, what is the difference between the heat transfer models presented in Chapter 2?
Using the transient calculation method from Chapter 5, the effect of different engine settings is numerically investigated. Therefore, a design of experiments, consisting of different operating conditions like varying air-fuel ratios, adjustments of ignition times, variations in the maximum fuel flow rate, and the change from a conventional to a prechamber spark plug, is presented. For all engine mappings, in-cylinder pressure measurements from the corresponding full load curve serve as input values. In addition, each of the mentioned variations is simulated

with all four presented heat transfer models, namely the approaches according to Woschni (Woschni, 1967), Hohenberg (Hohenberg, 1979), Eichelberg (Eichelberg, 1939), and Bargende (Bargende, 1991).
For comparison purposes, the engine settings are also experimentally investigated with the help of a four-cylinder engine. Therefore, racing simulations on an engine dynamometer with different race tracks and hybrid strategies are used. The measurement positions are the same as in Fig. 5.8 and 5.7. In this chapter, it is important to know that the cylinder individual behavior in Fig. 5.15 is also true for the other two cylinders. In summary, cylinder one and three, as well as cylinder two and four, behave thermodynamically similar. Therefore, these cylinder pairs are correspondingly numbered with one and two. In the following, using an underscore, the variable names also include the cylinder, or the engine block, number. For both, the experimental investigation, as well as the comparison of various heat transfer models within a numerical study, using several measurement points near the combustion chambers, a strictly statistical evaluation is used.

6.2 Method used in this chapter

For the transient temperature measurements, the engine mapping is systematically varied. With each characteristic setting, a race lap is replicated on an engine dynamometer. To overcome the thermal inertia, several laps are driven. Afterwards, with regard to the calculation method, these experimental results are compared with corresponding, transient simulations. Therefore, the four aforementioned heat transfer models, namely Woschni (Woschni, 1967), Hohenberg (Hohenberg, 1979), Eichelberg (Eichelberg, 1939), and Bargende, (Bargende, 1991) with Bargende et al. (2011), are integrated into the previous described method in Fig. 5.2. To evaluate the effect of different engine settings on component temperatures, their differences, in relation to the engine basic setting, are investigated. Therefore, mean values of the highly dynamic responses are compared. For the purposes of an expedient evaluation, a full load averaging is used, e.g., only full load periods are compared. This is physical meaningful, because the coasting conditions are mainly affected by the water temperature, its mass flow rate, and the air mass within the cylinder.

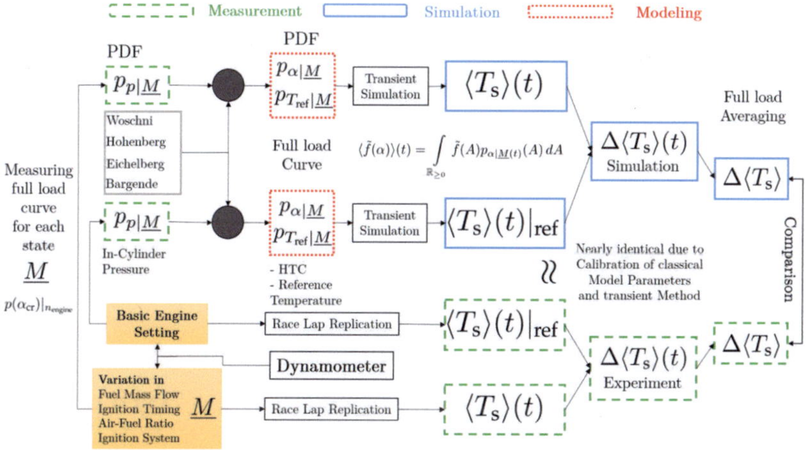

Figure 6.1: Method to investigate the influence of various engine settings on the transient, but cycle-averaged, temperature field $\langle T_\mathrm{s}\rangle(\boldsymbol{x}, t)$. Each engine setting is also dynamically simulated with different heat transfer models, and compared with corresponding experimental results.

These conditions are quite independent of engine settings. According to Fig. 5.2, at the beginning of each transient simulation, as a kind of engine characterization, a full load curve for each setting has to be measured. Therefore, crank angle resolved, in-cylinder pressure measurements for discrete engine speed points are necessary. An overview about the investigation and comparison method can be seen in Fig. 6.1.

6.3 Design of experiments

Fig. 6.2 shows schematically the experimental plan for the dynamic engine temperature measurements. It consists of single- and multi-stage realizations. For spark-ignition engines, there exist some constraints concerning mechanical and thermal loads, as well as the knocking level. Therefore, the independent factors cannot usually be varied without adjusting the others, resulting in important interactions. In addition, from the point of view of thermal efficiency, it does not always make sense to vary only one parameter. By way of example, while reducing the maximum fuel flow rate, one will always go along with increased ignition angles.

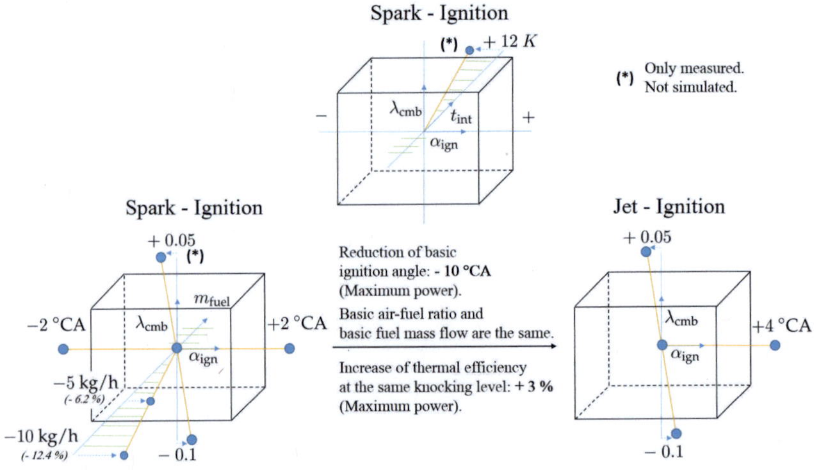

Figure 6.2: Design of experiments.

For the intake air, two different realizations are experimentally investigated: 24 °C and 36 °C. Due to the higher knocking tendency with higher fluid temperatures, the engine control unit adjusts the ignition angle. In this case, for an increase of the air temperature of about 10 K, the ignition angle is reduced by 1.8 °CA. The air-fuel ratio, as well as the fuel mass flow, is kept constant, resulting in an increase of the boost pressure of about 70 mbar. This is reached by increasing the exhaust pressure by the same amount. For this experiment, the ignition system consists of a conventional spark plug. In addition, a second ignition system, namely a prechamber ignition, is investigated. Subsequently, it will be called jet-ignition. To get the same distance to the individual knocking limits, the basic ignition angles of both combustion processes are individually set. This guarantees a fair comparison between both systems, and enables an ignition time variation in positive direction without destroying the engine. Moreover, there is no additional, significant thermal load due to knocking events, which could distort the measuring results. All other parameters like the air-fuel ratio and the fuel mass flow rate are the same. Given this basic engine setting, for both processes, an ignition time and air-fuel variation is investigated, resulting in a corresponding offset in the engine mapping. In addition, in case of the conventional spark plug system, a fuel mass flow variation is conducted. Here, in each case, the fuel flow under full load condition, measured in kilogram per hours, is set to a constant value. This corresponds to current sporting regulations of well-known racing series: Federation Internationale de l'Automobile - FIA (2018a) and Federation Internationale de l'Automobile - FIA (2018b). As already mentioned in Chapter 1, the main goal of these racing series is the increase of thermal efficiency, and, therefore, a simultaneous adjustment of the ignition angle is reasonable. However, due to high mechanical and thermal loads, only a reduction of the fuel flow is investigated. In case of the air-fuel variation, due to different knocking tendencies, an ignition time correction is necessary. Therefore, inversely related to the air mass, a reduction, or increase, of the ignition angle is simulataneously realized. Integrally, the combustion is always overstiochiometrical. For the ignition time, in case of the conventional spark plug, a two-stage realization with an offset of 2 °CA is chosen. With a lead of 4 °CA, the jet-ignition configuration is only tested with two different ignition angles.

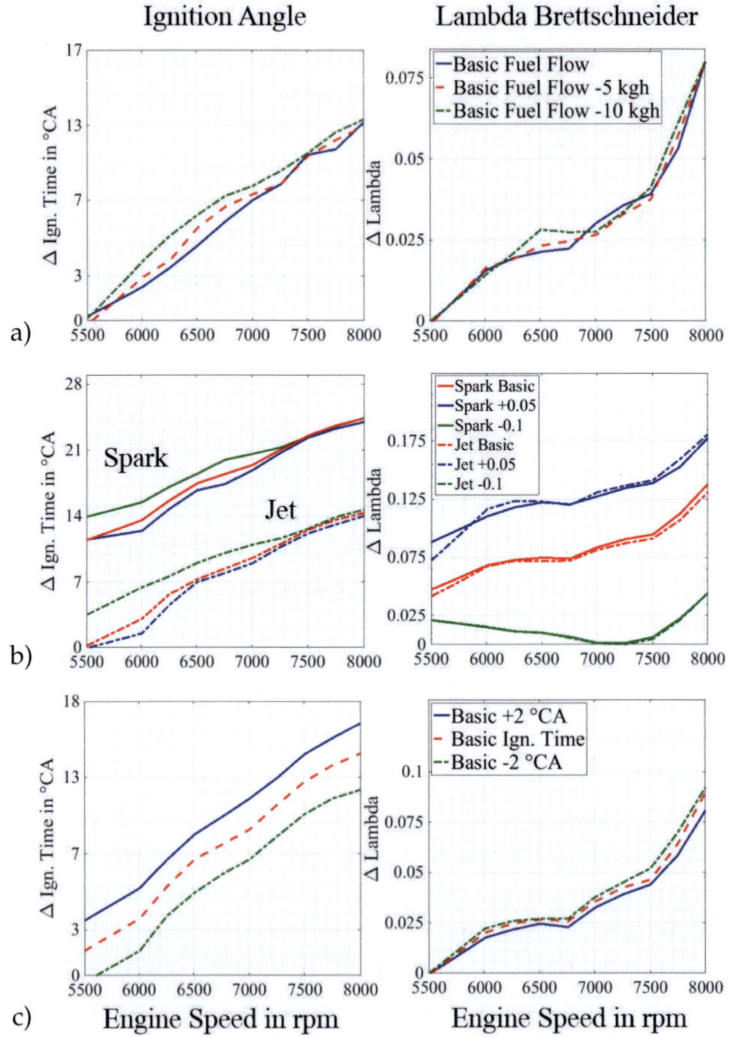

Figure 6.3: a) Fuel flow variation with conventional spark plug ignition. The air-fuel ratio is constant for each factor value, whereas the ignition time is adjusted for maximum thermal efficiency. b) Air-fuel variation with conventional spark plug and jet-ignition. The basic fuel flow is kept constant. Again, the ignition time is adjusted. c) Ignition time variation by the example of conventional spark plug ignition. The basic fuel flow is again constant.

An overview about the variations and the related full load settings are shown in Fig. 6.3. Due to the fact that dynamic temperature measurements are conducted, and that the race engine spends most of its time in coasting or full load conditions, only full load lines are visualized. As can be seen in Fig. 6.3 b), in case of the air-fuel variation, due to an increasing knocking tendency with lower engine speeds, larger ignition time adjustments are necessary. In addition, for the same reason, the constant offset in the basic air-fuel engine mapping, cannot be realized anymore for these low rotation rates. However, such low engine speeds do not often occur under dynamic race conditions. In Fig. 6.3 c), concerning the air-fuel ratio, a diverging trend can be observed in case of ignition time variations. This is due to some modeling errors for the dynamic exhaust pressure dependency of the lambda sensors used in the engine control unit. The plotted values in Fig. 6.3 c) are measured with a spatially separated, and markedly more accurate measurement equipment. The air-fuel ratio is calculated according to Brettschneider.

6.4 Results and discussion

6.4.1 Track characteristic and coasting strategy

As an example, Fig. 6.4 shows the influence of the race track and the sailing strategy. In this case, due to the fact that two different tracks, namely the Circuit de Barcelona-Catalunya and the Yas Marina Circuit in Abu Dhabi, are compared, the average values over the first 40 seconds are used. A change of full load condition from 54 % to 64 %, as well as a disappearance of a sailing cut off strategy, cause an average temperature increase in the range of 6 - 8 K, which corresponds to a reduction of the temperature gradient between solid and water in the order of 20 - 25 %. In this example, the simulated values are in good agreement with experimental results, and the differences are in the range of approximately one Kelvin.

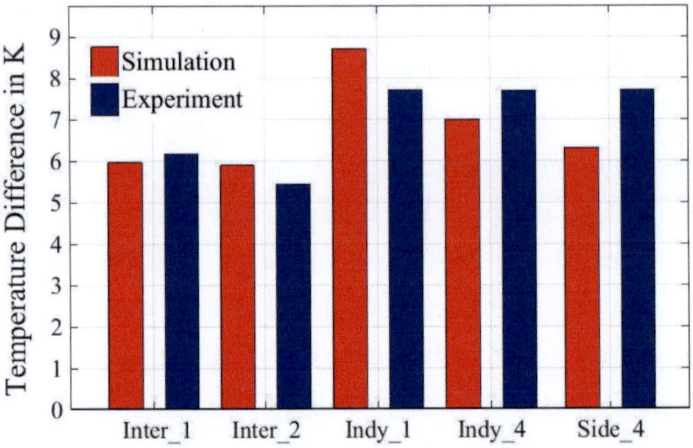

Figure 6.4: Temperature differences near the combustion chamber due to a change of the race track and the sailing strategy.

6.4.2 Fuel mass flow variation

According to Fig. 6.2, the fuel mass flow variation corresponds to a decrease of 6.2 % and 12.4 %. However, for the sake of clarity, temperature differences are referred to the second fuel flow rate, e.g., the middle one. In general, low temperature effects can be measured. The earlier arrival at the end of straight, due to a higher fuel mass flow rate, partly compensates the higher heat fluxes in transient engine conditions. In Fig. 6.5, the differences directly at the combustion chambers, within the cylinder heads, are shown. Similar to the stationary experiments in Fig. 3.11, all models, except the approach according to Eichelberg, show a comparable sensitivity. However, for the lowest fuel mass flow rate, this model is less accurate. Generally speaking, all models overestimate the effect of fuel mass flow variations. However, fuel flow variations in an HCCI engine are investigated in Broekaert et al. (2017). In this study, the models according to Woschni, Hohenberg and Bargende cannot capture the effect of the fuel mass flow rate on the maximum heat flux very well. In addition, also in case of an HCCI engine, Chang et al. (2004) studies the models according to Woschni and Hohenberg by means of a cumulative gross heat release error. In this case, a similar, apparent systematic error trend, which follows the load change, can be observed. Both studies, in contrast to the current results, present a too low model sensitivity for stationary engine conditions. A possible reason could be the difference in the combustion process. As can be seen in Fig. 6.6, a similar result can be found for the positions within the cylinder liner shoulders. In this case, in accordance with Fig. 5.16, two different sensitivities for the intake and exhaust side can be observed experimentally, as well as with the simulated values. Compared with experimental results, the sensitivity of the exhaust channel temperatures is slightly too large. Fig. 6.7 shows the comparison between simulated and measured temperature differences.

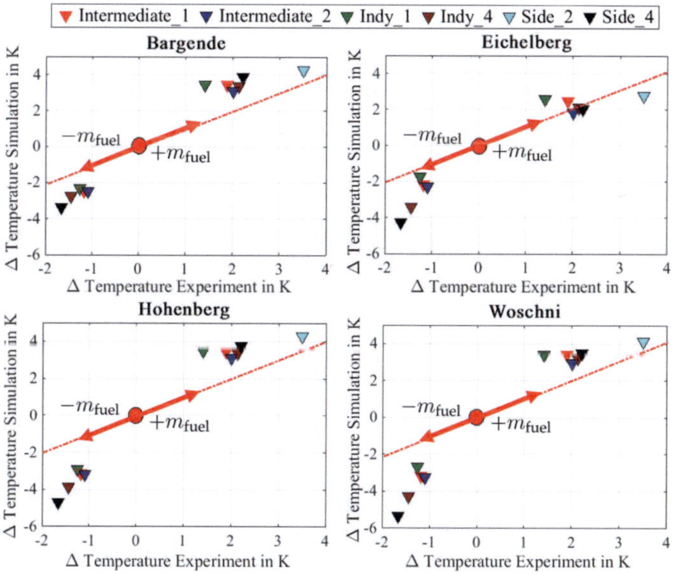

Figure 6.5: The effects of the fuel mass flow variation directly at the combustion chambers.

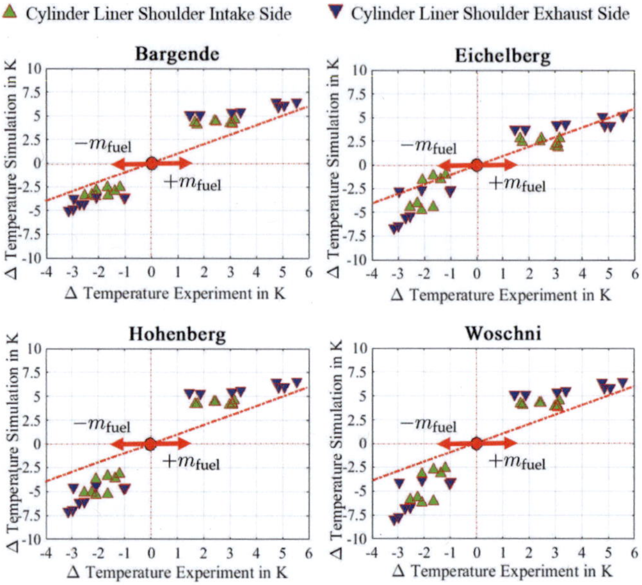

Figure 6.6: The effects of the fuel mass flow variation at the cylinder liner shoulders.

One possible explanation is that the reference temperature, according to Fig. 5.2, is the measured exhaust gas temperature under stationary engine conditions. Under dynamic conditions, the engine component temperatures are partly lower resulting in a larger temperature difference between gas and solid, e.g., a higher heat flux. That will decrease the effective exhaust gas temperature. In addition, the chemical kinetics needs time: the position for the stationary temperature measurement is located at the end of the exhaust pipe. Therefore, changes in the measured exhaust gas temperature do not influence effectively the driving temperature gradient at the beginning of the pipe. Obviously, the heat transfer models within the combustion chamber do not have a significant influence on exhaust channel temperatures. The different behavior for the two cylinder pairs is explained in Fig. 6.8, which shows the progression of the corresponding exhaust gas temperature.

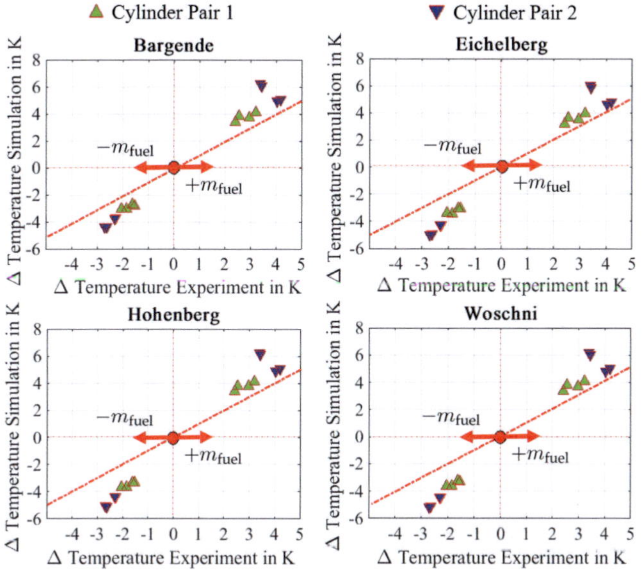

Figure 6.7: The effects of the fuel mass flow variation at the exhaust channels.

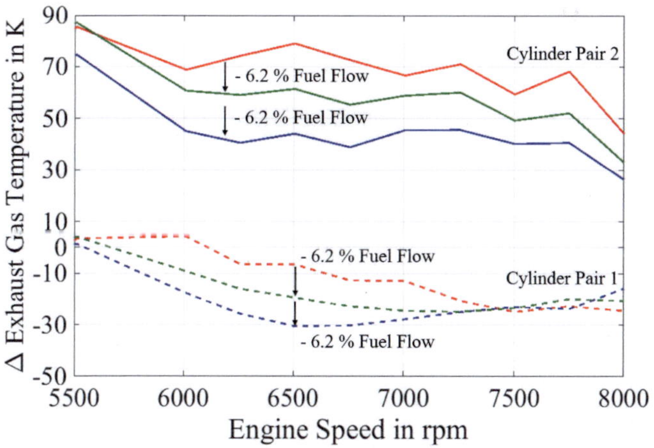

Figure 6.8: Change of the exhaust gas temperature during the fuel flow variation.

In accordance with the observed component temperature differences, cylinder pair two has a higher sensitivity to mass flow variations. As can be seen in Fig. 6.9, the combustion chamber models have an influence on the intake ring temperatures. However, the temperature differences are quite small under fuel flow variations. Remember that the hole drilling method, together with the corresponding position tolerances for the measurement positions, can affect the results. A comparison between Fig. 6.7 and 6.9 shows the higher sensitivity of the exhaust section, compared to the intake part.

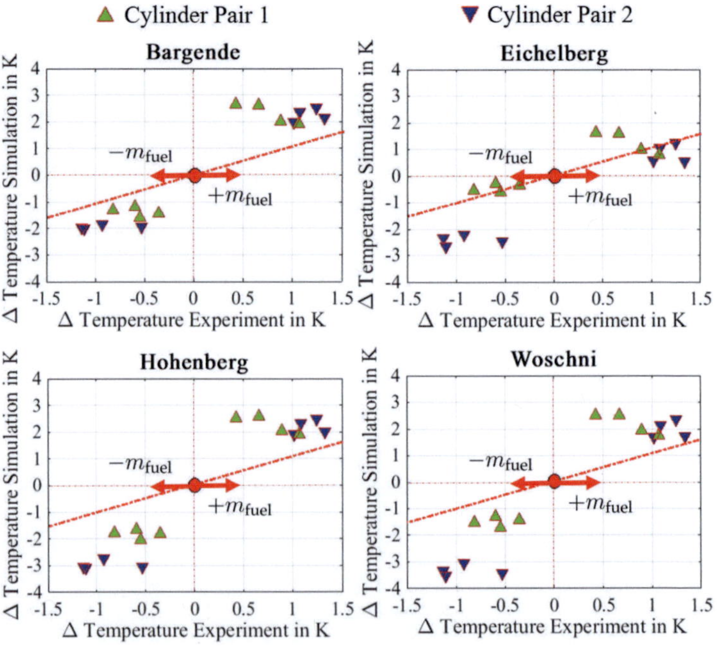

Figure 6.9: The effects of the fuel mass flow variation at the intake valve rings.

The unequal thermal behavior of cylinder pair one and two in Fig. 6.7 and 6.9 is caused by the already mentioned unsymmetrical ignition sequence and the resulting different volumetric efficiencies with unequal residual gas. Of course, the different air-fuel ratios in Fig. 5.15, including the corresponding differences in the fuel flows, have an influence on the intake and exhaust system. Intake channels are not significantly affected by fuel mass flow variations. In case of the exhaust valve rings, due to the already mentioned intrusive measurement method, resulting from the relative high thermal conductivity of the thermocouples, an evaluation is highly questionable. In summary, concerning fuel mass flow variations under dynamic engine conditions, the model according to Eichelberg seems to be the best choice for temperature predictions. In addition, it should be noted that the first reduction of 6.2 % can be simulated much better than the second reduction of 12.4 %. This factor level seems to be a too large difference to the calibrated engine state. The model according to Bargende, however, predicts, in accordance with experimental results, a more linear behavior for the complete variation range.

6.4.3 Lambda variation - conventional spark plug

In this experiment, due to some damaged piezoelectric sensors for the in-cylinder pressure indication, only a reduced number of measurement positions can be used. In addition, in case of the leaner combustion, the perturbation does not allow a clean evaluation. Therefore, only the effects of a richer mixture setting is investigated. As can be seen in Fig. 6.10, compared with the exhaust channels and valve rings, the intake section shows a higher sensitivity to air-fuel variations. Compared with fuel mas flow variations, the effect behaves exactly the other way round. Due to a richer combustion with a constant fuel mass flow, less air flows through the inlet system resulting in higher intake channel and ring temperatures. Especially in case of the intake valve rings, the hotter gas temperatures within the combustion chamber result in an additional heat flux towards the intake system. However, the simulated differences are not that high.

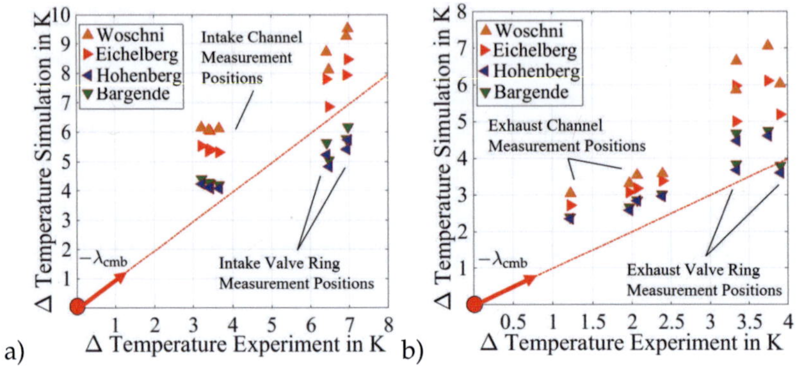

Figure 6.10: a) The effects of the lambda variation at the intake system. b) The effects of the lambda variation at the exhaust system.

Again, in case of the cylinder liner, a clear difference concerning the intake and exhaust side can be observed. Temperature differences, together with one measurement point near the combustion chamber, e.g., Indy at cylinder one, are shown in Fig. 6.11. Because of a higher sensitivity of the intake channels and valve rings, compared with the exhaust system, the effect on the intake side is more pronounced.

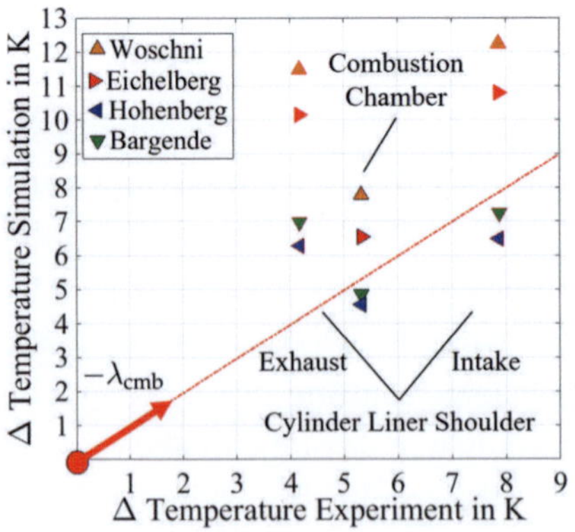

Figure 6.11: The effects of the lambda variation at the combustion chamber and cylinder liner shoulders.

This difference can also be seen in the simulation. However, it is not as strong as in the measurements. The clear overestimation of all models for the exhaust side of the liners, as well as for the exhaust channels and rings, can also be explained by the same arguments than in case of the fuel flow variation. In summary, the two models according to Bargende and Hohenberg have the highest prediction accuracy concerning the temperature increase towards a more stoichiometric combustion. In contrast to the stationary results in Fig. 3.13, the model according to Woschni clearly overestimates the effects under dynamic engine conditions.

6.4.4 Ignition time variation - conventional spark plug

Compared with the two preceding variations, different ignition times cause small temperature effects. As can be seen in Fig. 6.12,

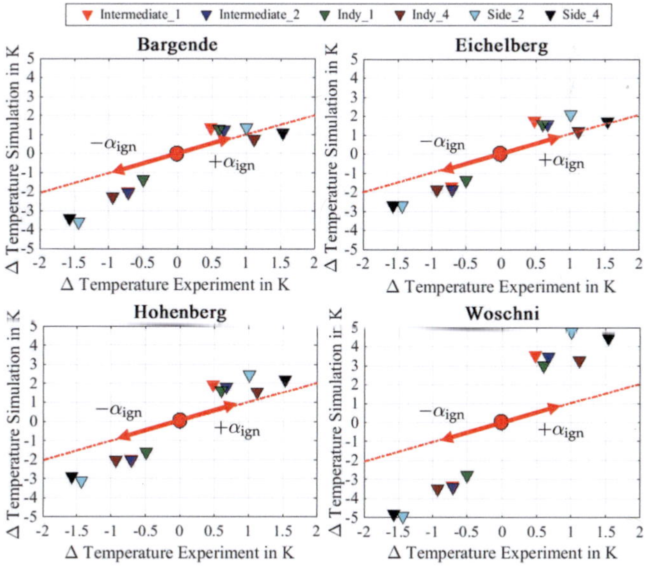

Figure 6.12: The effect of the ignition time variation at the combustion chambers.

the model according to Eichelberg gives a small derivation of about one Kelvin. However, in case of a positive ignition angle variation, the model according to Bargende gives the smallest errors at all. Similar to the fuel mass flow variation, laterally oriented points near the combustion chamber react more sensitively.

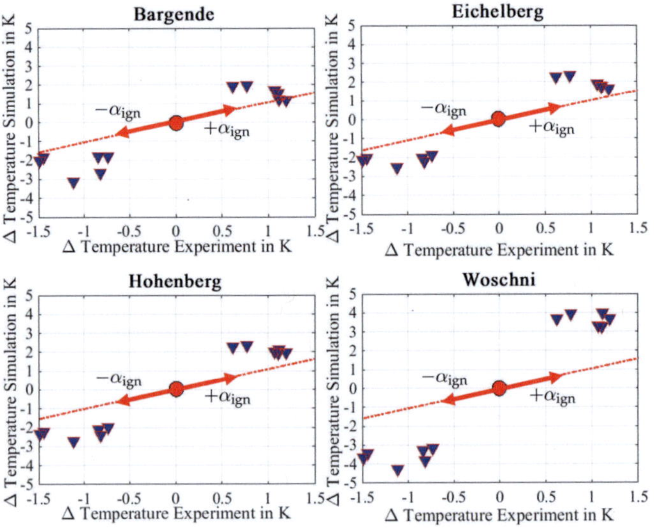

Figure 6.13: The effect of the ignition time variation at the intake rings.

As shown in Fig. 4.9 b), the small temperature change can be explained by the fact that only the peak pressure, and, therefore, the heat transfer coefficient with the gas temperature, rises significantly with larger ignition angles. For cycle averaged component temperatures, however, the complete engine cycle is relevant. In accordance with experimental results, an approximately linear behavior for the multi-stage factor variation can only be simulated with the model according to Hohenberg and Woschni. However, the last one clearly overestimates the sensitivity. This is in accordance with Fig. 3.9. A similar trend can be seen in Fig. 6.13 for the intake valve rings. However, the temperature changes are quite small. For the exhaust rings and the intake channels, no real tendency can be observed. However, in case of the exhaust channels, the reduction of the thermal efficiency with smaller ignition angles causes an energy transfer to the exhaust gas enthalpy, and, therefore, the gas temperature raises.

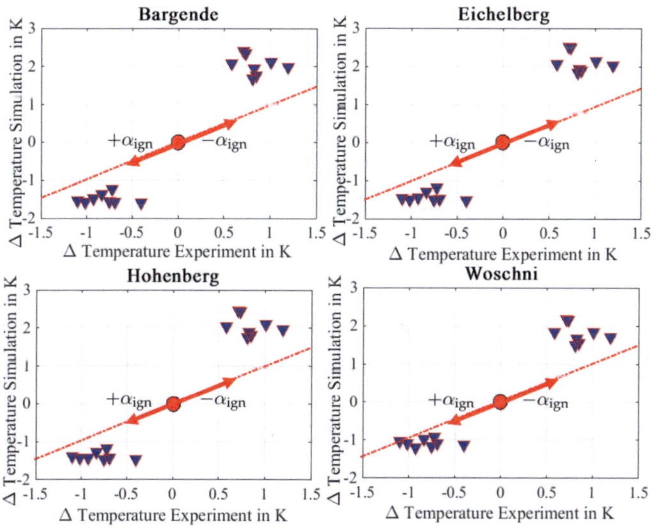

Figure 6.14: The effect of the ignition time variation at the exhaust channels.

As can be seen in Fig. 6.14, slightly higher exhaust channel temperatures are the result. Again, due to the aforementioned error sources, the simulation overestimates this effect. In turn, this trend is quite independent of the heat transfer models within the combustion chamber. For larger ignition angles, compared with the exhaust side, this cooling effect in the exhaust channels causes a higher temperature increase at the intake side of the cylinder liner shoulders. Fig. 6.15 shows some results: again the models according to Bargende and Eichelberg seem to be the best choice. The different sensitivities for the intake and exhaust side can also be recognized in the simulation.

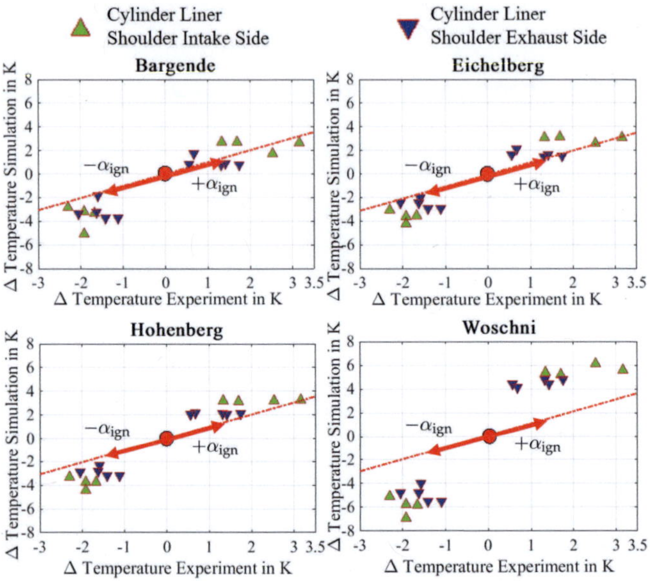

Figure 6.15: The effect of the ignition time variation at the cylinder liner shoulders.

6.4.5 Ignition process - conventional spark plug versus jet-ignition

Again, due to some conspicuous features of the piezoelectric sensors for the in-cylinder pressure indication, only a reduced dataset is shown. As can be seen in Fig. 6.16, in case of the combustion chamber and the cylinder liner shoulders, the jet-ignition causes higher component temperatures. Therefore, despite of a higher thermal efficiency, the heat loss for jet-ignition engines is higher. Obviously, towards a perfect Otto cycle, the increase in efficiency results from a faster combustion, e.g., a larger peak pressure. In summary, the models according to Eichelberg and Hohenberg seem to be the best choice. However, the temperature increase in the range of two Kelvin is not that high. For the intake system, consisting of channels and valve rings, the temperature is not much affected. The increase is in the range of one and three Kelvin.

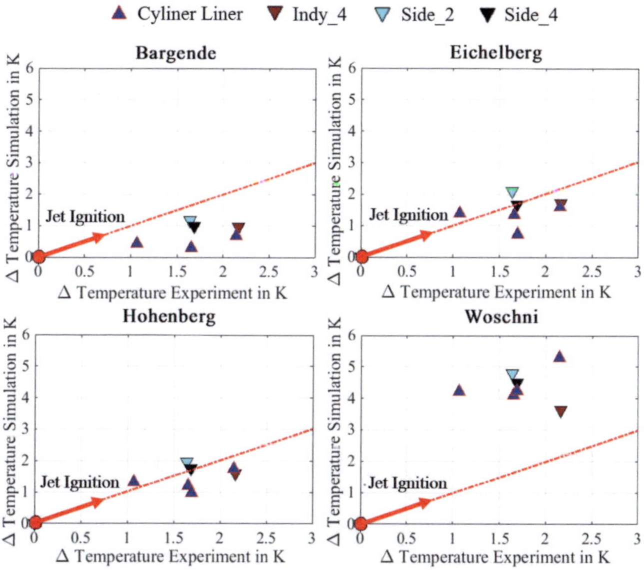

Figure 6.16: The effect of the ignition variation at the combustion chamber and the cylinder liner shoulders.

The predicted temperature changes, together with the experimental ones, can be seen in Fig. 6.17. Clearly, due to the greater proximity to the combustion chamber, the valve rings show a higher sensitivity. Again, the approaches according to Eichelberg and Hohenberg seem to be the most suitable models. In accordance with the combustion chamber mea surements in Fig. 6.16, the model according to Bargende underestimates the effect of jet-ignition, whereas the model by Woschni repetitively overestimates the changes. Concerning the exhaust valve rings, there exist an extremely small temperature change, mainly less than one Kelvin, caused by the increased heat transfer within the combustion chamber on the one hand, and due to a cooler exhaust gas temperature on the other hand. Of course, the last one is the result of the higher thermal efficiency. Analogously, the simulated changes can be considered as noise. Similarly, the exhaust channels are mainly affected by the cooler exhaust gas temperatures resulting in lower channel temperatures.

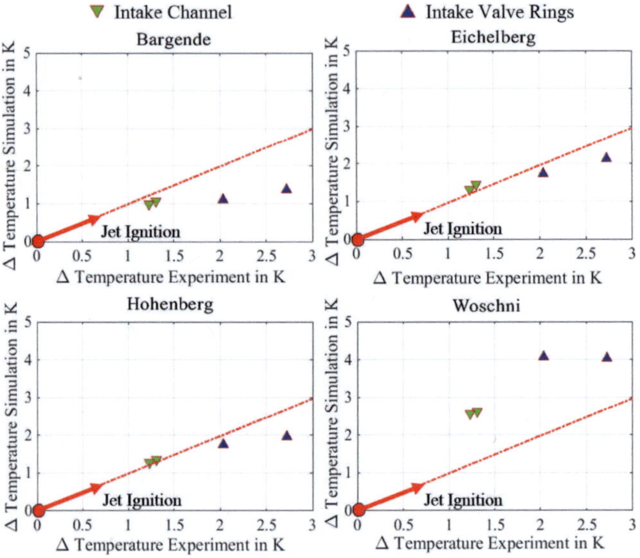

Figure 6.17: The effect of the ignition variation at the intake system.

Again, the simulated values are, therefore, independent of the heat transfer models within the combustion chamber. However, analogously to the valve rings, these changes are quite small: the largest, measured temperature change is 0.5 K. Its corresponding simulated value is, using the model according to Bargende, 0.63 K.

6.4.6 Ignition time variation - jet-ignition

Concerning ignition time variations under jet-ignition, one key observation in Fig. 6.18 is that the temperature differences near the combustion chamber are very similar to changes under a conventional spark plug ignition. Remember that in this case the change in the ignition angle is 4°CA.

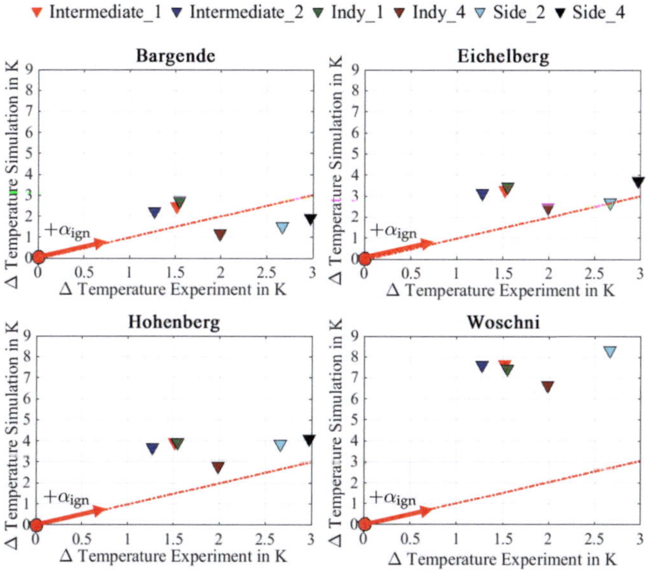

Figure 6.18: The effect of the ignition time variation at the combustion chambers under jet-ignition.

As can be seen in Fig. 6.18, in accordance with Fig. 6.12, the model according to Bargende can best predict temperature differences due to ignition time variations. Again, the model by Woschni totally overestimates the ignition time effect. For jet-ignition, in comparison with Fig. 6.15, differences between the intake and exhaust side of the cylinder liner shoulders cannot be observed anymore. The results can be seen in Fig. 6.19. This can be explained by the fact that, under ignition time variations, the change of the exhaust gas temperature is, in comparison with a conventional spark plug ignition, not that high: at nominal speed, in case of the conventional system, the gas temperature increase, caused by a reduction of the ignition angle of about four degrees, is 33 K. For the jet-ignition, the difference is only 12 K. Both values are cylinder averaged. However, in case of the cylinder liner shoulders, there exists another interesting effect, which cannot be observed to the same extent during the other variations.

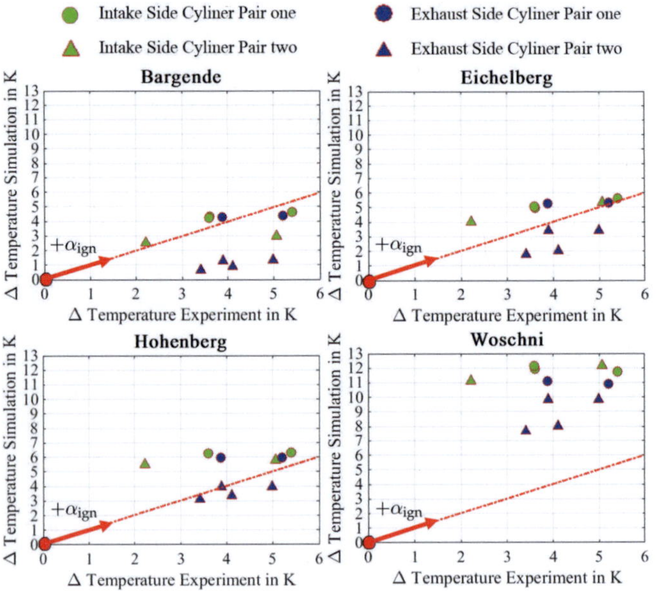

Figure 6.19: The effect of the ignition time variation at the cylinder liner shoulders under jet-ignition.

Probably, because of the aforementioned unsymmetrical ignition sequence for the investigated engine and the resulting different volumetric efficiencies with unequal residual gas in the two cylinder pairs, all four models show two different sensitivities for these pairs. However, the experimental results do not show this pronounced trend. Concerning component temperatures, it seems as if all models overestimate the effect of the different pressure curves. For both pairs, the same kind of pressure transducer is used. In addition, there is no hint that one of these transducers are damaged. This can be ensured by evaluating the difference between the measured, effective torque and the indicated mean pressure. In case of the cylinder liners, this makes a suitable model choice difficult.

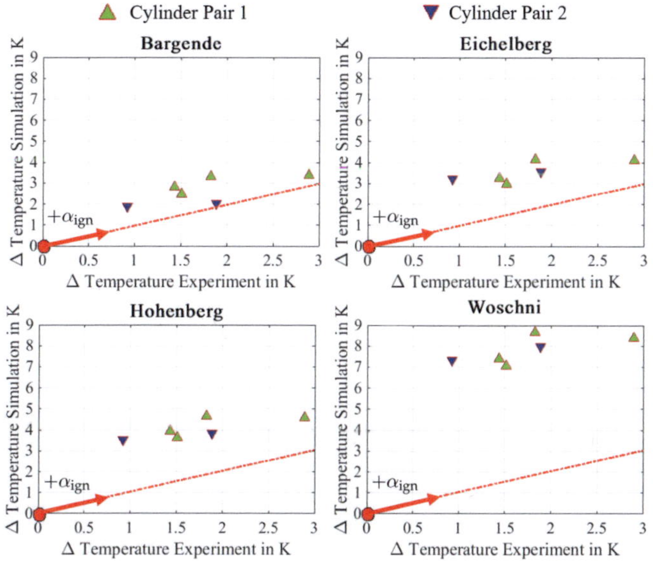

Figure 6.20: The effect of the ignition time variation at the intake valve rings under jet-ignition.

In addition, the scattering character for these measuring positions is significantly pronounced. As can be seen in Fig. 6.20, this wrong tendency cannot been observed anymore in case of the intake valve rings. For larger ignition angles, one can see slightly higher temperature changes at the cylinder pair one for both the measured and simulated values. However, in accordance with Fig. 6.13, the temperature differences are quite small at all. Similar to the ignition time variation with a conventional spark plug, no real tendency can be observed for the exhaust rings and the intake channels. However, comparing the temperatures of the exhaust channels in Fig. 6.21 and 6.14, one can observe a similar behavior. Although the exhaust gas temperature does not decrease that significantly, in case of jet-ignition, it is remarkably that the channel temperatures decrease to the same extent.

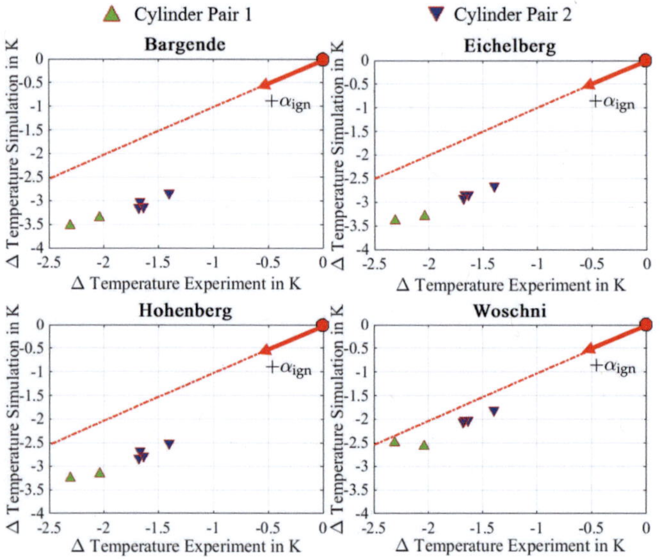

Figure 6.21: The effect of the ignition time variation at the exhaust channels under jet-ignition.

Again, due to the already discussed error sources, the simulation method overestimates this effect with one or a half Kelvin, depending on the heat transfer model. However, it would be a fallacy to conclude that the model according to Woschni is the best choice. In this case, the too strong cooling is partially offset by the overestimated heat transfer within the combustion chamber.

6.4.7 Lambda variation - jet-ignition

Fig. 6.22 shows the resulting temperature differences for the measurement positions near the combustion chamber. Interestingly, a lambda increase of 0.05 shows a significantly larger sensitivity, compared to an analogous decrease of 0.1.

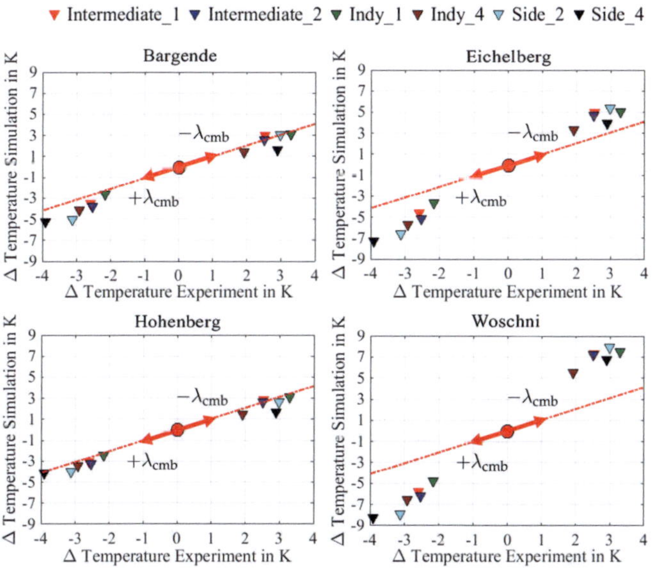

Figure 6.22: The effects of the lambda variation at the combustion chambers under jet-ignition.

Obviously, the models according to Bargende and Hohenberg have the highest prediction accuracy. In comparison with Fig. 6.11, a decrease of the air-fuel ratio by the same amount results in a lower sensitivity, e.g., the measurement positions near the combustion chamber do not reduce their temperatures by the same amount. Fortunately, the simulation can also predict this observation. As can be seen in Fig. 6.23, concerning the cylinder liner shoulders, the changes are, however, more similar. With regard to the exhaust channel temperatures, the influence of the air-fuel ratio is not that high. For the underlying engine, each factor level is globally overstiochiometrical, and the thermal efficiency increases with lower λ_{cmb} - values. On the other hand, however, the effective, thermal capacity also decreases. Therefore, the exhaust gas temperature is not very affected.

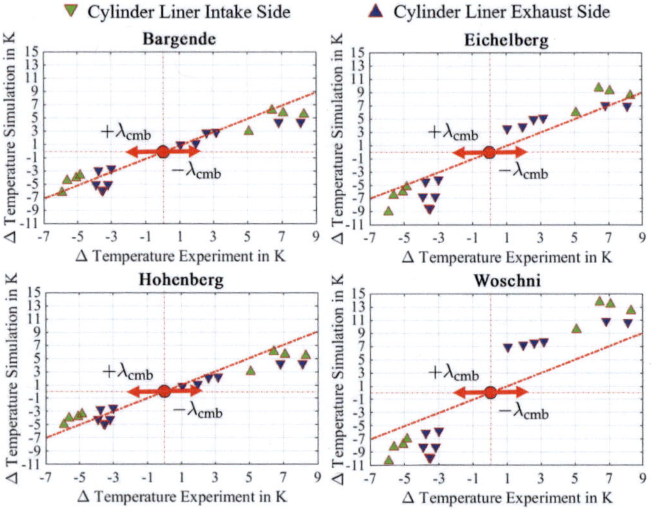

Figure 6.23: The effects of the lambda variation at the cylinder liner shoulders under jet-ignition.

In addition, the aforementioned, cylinder individual effects make the tendency less transparent: the exhaust channel temperatures should be considered in conjunction with each cylinder. Again, the heat transfer models within the combustion chamber do not strongly affect these temperatures. Therefore, as an example, only the model according to Hohenberg is shown and compared with some experimental results in Fig. 6.24. For the exhaust valve rings, the situation is even more complex. In this case, the heat transfer within the combustion chamber, as well as the exhaust gas temperature, will define the temperatures. The results for all four cylinders, including the leaner and richer configuration, are shown in Fig. 6.25. As expected, the models according to Hohenberg and Bargende can best predict the temperature differences. In accordance with Fig. 6.22, a richer combustion results in higher temperatures.

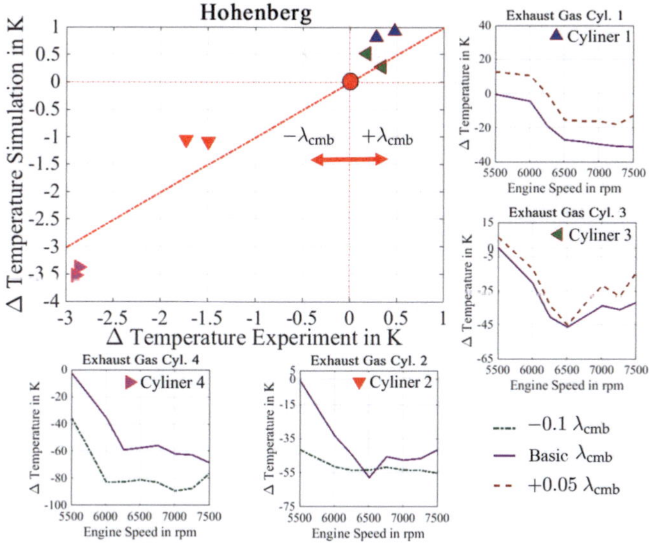

Figure 6.24: The effects of the lambda variation at exhaust channels under jet-ignition.

However, the sign of the temperature change is also affected by the exhaust gas temperature, and, therefore, it is a cylinder individual result. The temperature changes for the intake system are shown in Fig. 6.26. Due to their close proximity to the combustion chamber, in accordance with Fig. 6.10, the valve rings react much more strongly than the channels. This behavior can also be seen in the simulation, and, again, the model according to Hohenberg can best predict the corresponding temperature changes. Compared to a conventional spark-ignition, the temperature sensitivity under jet-ignition is, however, lower. The same factor level results in a change by one Kelvin, in case of the intake channels, and up to five Kelvin at the valve rings. Under spark-ignition, the changes are in the range of four and seven Kelvin. In accordance with the increase of the intake channel and valve ring temperatures for a richer combustion, the intake side of the cylinder liner shoulders correspondingly increases somewhat stronger than the exhaust side. The simulated values, together with the measured temperature differences, are shown in Fig. 6.23.

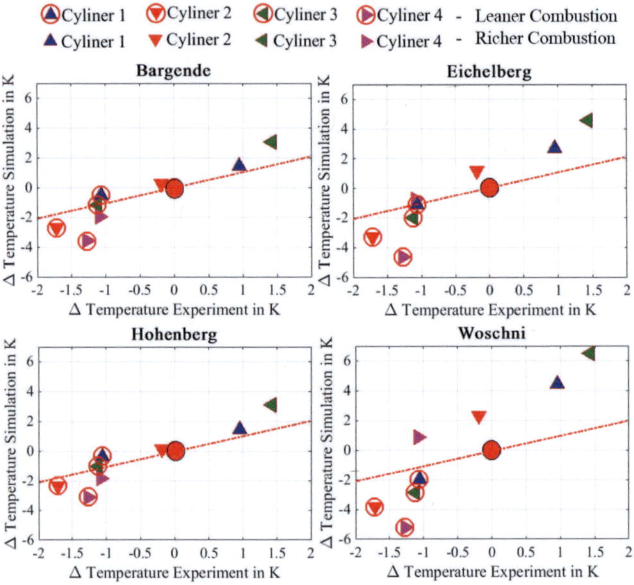

Figure 6.25: The effects of the lambda variation at the exhaust rings under jet-ignition.

6.4.8 Intake air temperature variation

Although the ignition angle is reduced, and the effective power of the engine decreases, higher intake air temperatures result in higher component temperatures near the combustion chamber. However, this increase is very small for most of the measurement positions. The two values in Fig. 6.27 represent the maximum range. By way of example, the other positions within the cylinder head, near the combustion chamber, increases in the order of 0.1 K. Analogously, the other measurement positions within the cylinder liner shoulders show an increase between 0.5 and 1.5 K. Unfortunately, during the air temperature variation, the pressure indication measurement systems for the individual cylinders show a conspicuous tendency, and, therefore, a clean comparison cannot be ensured anymore.

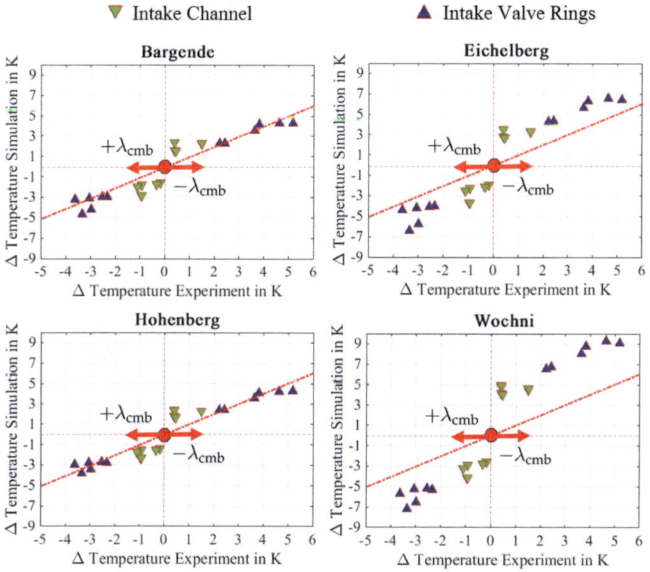

Figure 6.26: The effects of the lambda variation at the intake system under jet-ignition.

Due to the very small measured differences, the main effects are overshadowed to a considerable extent by this failure, and by the measurement errors for the model input parameters. Therefore, in this special case, simulations are very questionable, and have not been conducted. For the sake of completeness, however, the measured values are given: For the intake channels, the temperature increase is in the range of 1 and 3 K. For the exhaust channels, due to a lower thermal efficiency and higher exhaust gas temperatures, the increase is between 1 and 1.4 K. Analogously, the temperatures within the intake and exhaust valve rings increase by 0.5 - 1.5 K and 0.6 - 1.2 K.

6.5 Summary and conclusion

To summarize the different temperature effects of the investigated variations, two different engine components, namely the cylinder head and the cylinder liner, made from aluminium and steel, are evaluated. In summary, for the variations in Fig. 6.2, the maximum temperature range which can be measured for the positions near the combustion chamber is shown in Fig. 6.27 a).

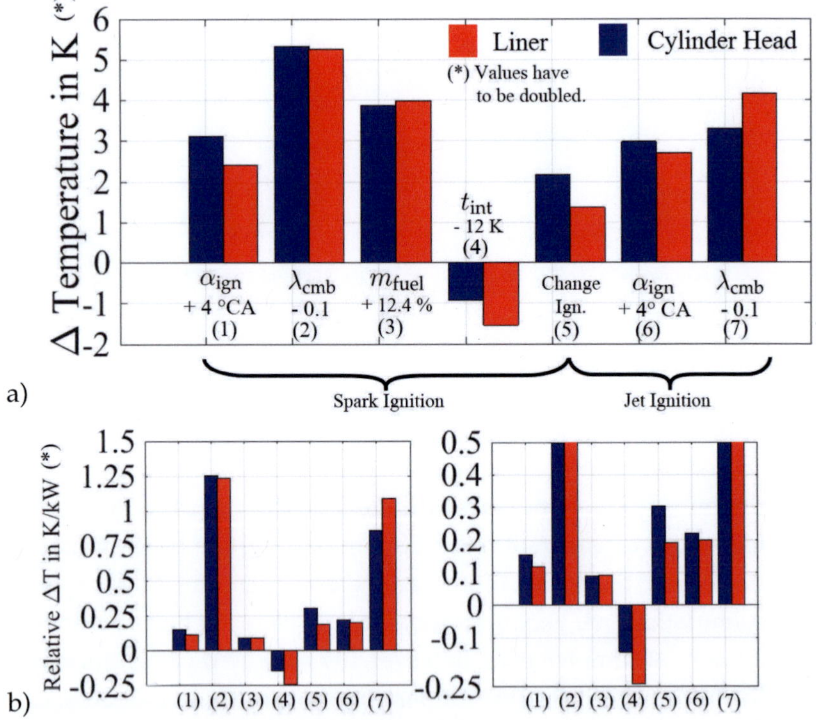

Figure 6.27: a) Maximum range of temperature near the combustion chamber for different kind of variations. Measured values for the cylinder head, in blue, and for the cylinder liner, in red, are shown. For the last one, the given values have to be doubled. The factor levels are given below. b) Temperature differences in relation to the corresponding, effective power increase.

In addition, the values for the cylinder liner shoulders are given. In relation to the different variations, despite diverse materials and dynamic responses, both components behave qualitatively very similar, and, therefore, the comparison should be reasonable. By the example of the cylinder head, with more than five Kelvin, variations in the air-fuel ratio under spark-ignition result in the largest temperature differences. Therefore, beside water temperature changes, shown in Fig. 5.20, they offer the largest control leverage. Under jet-ignition, the values are two Kelvin lower. Ignition time adjustments give, for both ignition systems, a temperature change in the range of three and six Kelvin. However, the combustion chamber is only little affected by a pure change of the ignition system. Interestingly, due to the cooler fluid temperature, a decrease of the inlet air temperature results, despite a higher effective power, in colder component temperatures. The investigated engine has a power density in the range of 180 kW per liter. With regard to the corresponding change of effective power, measured at nominal speed, the relative temperature differences can be seen in Fig. 6.27 b). In case of spark- and jet-ignition, relative temperature differences of about one Kelvin per kilowatt, resulting from variations in the air fuel ratio, are clearly the largest. The other variations result in much lower values in the range between 0.1 and 0.3 Kelvin per kilowatt. Again, in case of inlet air temperature changes, the sign is different.
To sum up the model comparison, the empirical expectation value $\tilde{E}$ and standard deviation $\tilde{\sigma}_E$ of the error are compared for each factor:

$$\tilde{E} = \frac{1}{K} \sum_{k=1}^{K} E_k, \tag{6.1}$$

$$\tilde{\sigma}_E = \sqrt{\frac{1}{K-1} \sum_{k=1}^{K} \left(E_k - \tilde{E}\right)^2}. \tag{6.2}$$

For a measurement position k, the error between simulation and experiment is described by E_k. K is the complete number of measurement positions. For multi-stage realizations, each factor level is considered separately, but are summarized in one value for the expectation value and the standard deviation. Therefore, the factor level is also given below. Positive errors mean an overestimation of the underlying effect. Due to the high model dependency, positions near the combustion chamber, e.g., the cylinder head and the cylinder liner, are chosen. Due to some damaged thermocouples, there exist an unequal number of measurement positions for each factor. Two-stage realization experiments have nearly twice the number of positions as one-stage variations. In summary, for transient temperature simulations, it must be mentioned that the heat transfer model within the combustion chamber has to be defined individually in each case. For each factor variation, a comparison between the four investigated models is shown in Fig. 6.28. Therefore, under the assumption of a Gauss distribution

$$p_E = \frac{1}{\sqrt{2\pi\tilde{\sigma}_E^2}} \exp\left(-\frac{\left(E - \tilde{E}\right)^2}{2\tilde{\sigma}_E^2}\right), \tag{6.3}$$

for the error E, the different probability density functions are presented. For the expectation value and standard deviation, the empirical estimators from Eq. (6.1) and (6.2) are used. Their values are also given in Table 6.1. It can be seen that, for both ignition systems, the variation of the air-fuel ratio can be best simulated by using the model according to Bargende. Despite of large temperature changes, the expectation value is nearly zero. Similarly, for ignition time adjustments, the Bargende model seems to be the best choice. Compared with the jet-ignition, in case of the spark-ignition, there exists a slight drift towards an overestimation during the ignition time adjustments. In summary, the relative evaluation of the different heat transfer models is similar for jet- and spark-ignition. As already mentioned, each model overestimates the effect of fuel flow variations. In this case, the model according to Eichelberg gives the best expectation value. However, the standard deviation is quite large. Using the models according to Eichelberg or Hohenberg, good results can be achieved for the change from spark- to jet-ignition. The error expectation value, as well as the standard deviation, is quite small. Interestingly, for all variations, the model according to Woschni overestimates significantly each effect.

	Factor Level	Bargende	Eichelberg	Hohenberg	Woschni
$\tilde{E}$	Jet - $\Delta\lambda_{\text{cmb}} = 0.1$	-0.14 K	2.02 K	-0.55 K	4.25 K
$\tilde{\sigma}_E$	Jet - $\Delta\lambda_{\text{cmb}} = 0.1$	1.24 K	1.27 K	1.08 K	1.19 K
$\tilde{E}$	Jet - $\Delta\alpha_{\text{ign}} = 4$ °CA	-0.81 K	0.45 K	1.25 K	6.22 K
$\tilde{\sigma}_E$	Jet - $\Delta\alpha_{\text{ign}} = 4$ °CA	1.60 K	1.30 K	1.30 K	1.44 K
$\tilde{E}$	Spark - $\Delta\lambda_{\text{cmb}} = 0.1$	-0.02 K	2.02 K	-0.33 K	3.10 K
$\tilde{\sigma}_E$	Spark - $\Delta\lambda_{\text{cmb}} = 0.1$	1.27 K	1.51 K	1.26 K	1.58 K
$\tilde{E}$	Spark - $\Delta\alpha_{\text{ign}} = 2$ °CA	0.85 K	0.94 K	1.22 K	3.23 K
$\tilde{\sigma}_E$	Spark - $\Delta\alpha_{\text{ign}} = 2$ °CA	0.77 K	0.61 K	0.52 K	0.70 K
$\tilde{E}$	Spark - $\Delta m_{\text{fuel}} = 12.4\,\%$	1.61 K	0.92 K	2.25 K	2.26 K
$\tilde{\sigma}_E$	Spark - $\Delta m_{\text{fuel}} = 12.4\,\%$	0.68 K	1.13 K	0.86 K	1.07 K
$\tilde{E}$	Spark - Jet	-0.94 K	-0.20 K	-0.22 K	2.24 K
$\tilde{\sigma}_E$	Spark - Jet	0.53 K	0.44 K	0.39 K	0.78 K

Table 6.1: Error expectation value $\tilde{E}$ and standard deviation $\tilde{\sigma}_E$ for various factor levels and different heat transfer models.

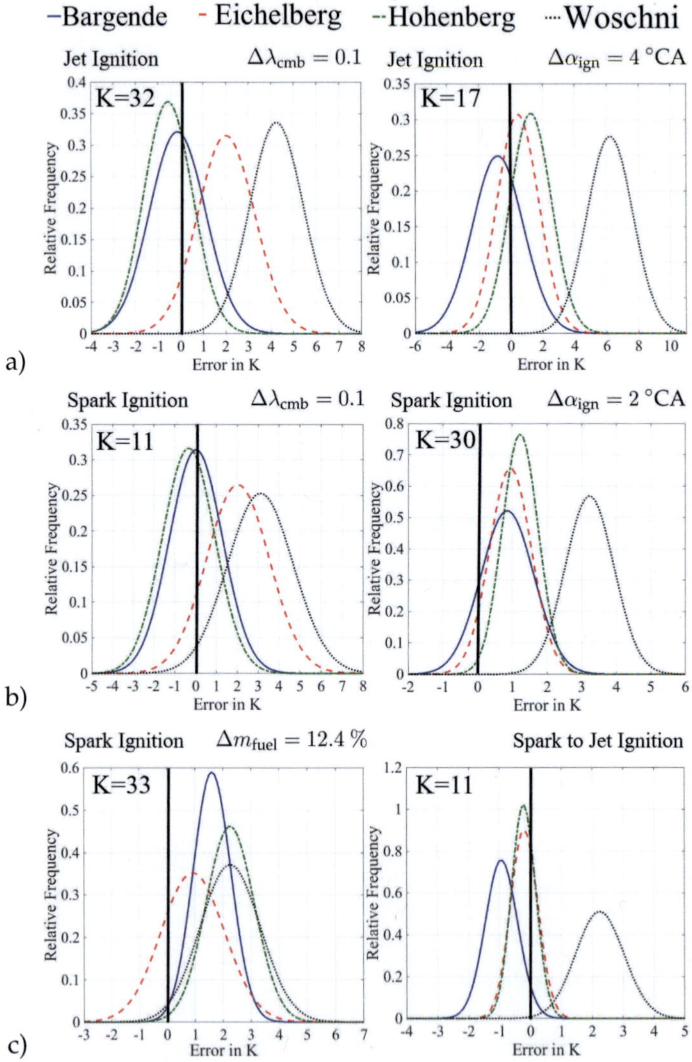

Figure 6.28: a) Model comparison for variations in the air-fuel ratio, as well as in the ignition angle under jet conditions. b) Model comparison for variations in the air-fuel ratio, as well as in the ignition angle under spark conditions. c) Model comparison for variations in the fuel mass flow, as well as for a change of the ignition system. For the statistical calculation, the underlying total number of measurement positions is given by K.

Chapter 7

Influence of various engine settings on the lifetime of engine pistons under dynamic operating conditions

7.1 Outline of the chapter

Race engines undergo highly transient thermal and mechanical loads during a race lap. This includes temperature fluctuations in the frequency range of 0.1-1 Hz, as well as strongly scattering pressure cycles in the order of 100 Hz. In addition, resulting from different engine speeds, there exist varying inertia loads. Therefore, the research question can be formulated as follows: with regard to high cycle fatigue, which damage results from different constellations of transient thermal and mechanical loads during a race lap? In particular, resulting from unequal temperature fields and pressure cycles, how do various engine settings differ from one another?

With the help of a new statistical method, due to the non-linearity of damage, cyclical fluctuations in the pressure curves for fixed engine states $\underline{M}$ are considered. Together with the investigation of various damage combinations of upper and lower stresses, resulting from different crank angle positions with changing pressure and side force loads, various system responses can be derived, and visualized with specially

defined Peak-Pressure Woehlerlines and conventional damage survivor functions. Therefore, a dynamic engine damage map is developed which combines different constellations of transient thermal and mechanical loads. Due to high-efficient operating points, and a high level of excess air in modern motorsport applications, the temperature level of the engine components is much less. In the following, cylinder two is chosen for the fatigue investigations. The maximum piston temperature at the end of straight is in the range of 250 °C. In addition, the dynamic behavior under race conditions results in relative low temperature amplitudes, and, therefore, no macroscopic, inelastic strains occur in many engine locations. However, the failure mechansims due to high cycle fatigue can be very pronounced.
With the help of the aforementioned damage maps, different engine settings, visualized in Fig. 6.2, are investigated and compared specifically for racing conditions. Therefore, the corresponding, different pressure cycles and dynamic temperature fields from Chapter 6, are used as boundary conditions.

7.2 Method used in this chapter

Using Abaqus as FE solver, various FEM studies with subsequent fatigue investigations are performed for an engine piston. An overview about the method is shown in Fig. 7.1. According to Fig. 6.2, for each engine configuration, including variations in the ignition system, ignition angle, air-fuel ratio, and the fuel mass flow, different transient, thermally stable temperature fields are mapped from the FVM model onto a tetra-hexa FEM Mesh, and impinged with resultant mechanical loads as stress boundary conditions. For the FEM model, a tight fit is realized between the piston hub and an additional ring. The subsequent assembly processing is modeled as a location-dependent clearance card for the bolt-ring contact. According to realistic contact patterns, the typical trumped shape of the bolt, as well as the detailed piston skirt form, are modeled. For the purposes of a dynamic analysis, three meaningful snapshots of the transient temperature simulation and three meaningful engine speeds are chosen to calculate damage diagrams with nine bases for various piston positions. The definition of "meaningful" is described in Fig. 7.2. The maximum piston temperature serves as a trigger criterium for the snapshots. Therefore, for a given race track, the global minimum and maximum, including their mean, are chosen. Analogously, the engine speed under full load conditions is discretized uniformly. The temperature fields are gained by using the calculation method from Chapter 5 and the most suitable, setting individual, heat transfer model from Fig. 6.28.

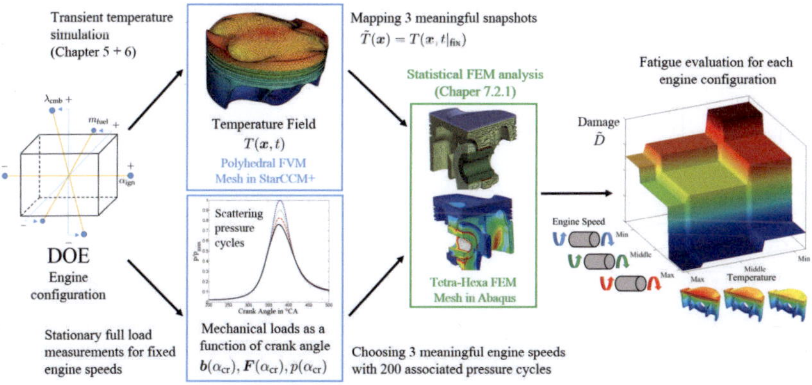

Figure 7.1: Overview about the numerical high cycle fatigue investigation of pistons. The thermal and mechanical loads depend on the various engine settings.

For each engine speed, the pressure curves are measured by a pressure indication measurement system, and the resulting acceleration $\boldsymbol{b}$ and piston side forces $\boldsymbol{F}$ are calculated with the help of the well-known engine kinematics. The first one is implemented as a force per unit volume, whereas the second one is transmitted on the piston skirt with a a cylindrical, rigid surface. However, due to the scattering nature of the gas pressure p and the corresponding lateral force $\boldsymbol{F}$ during the burning phase, a statistical analysis is necessary.

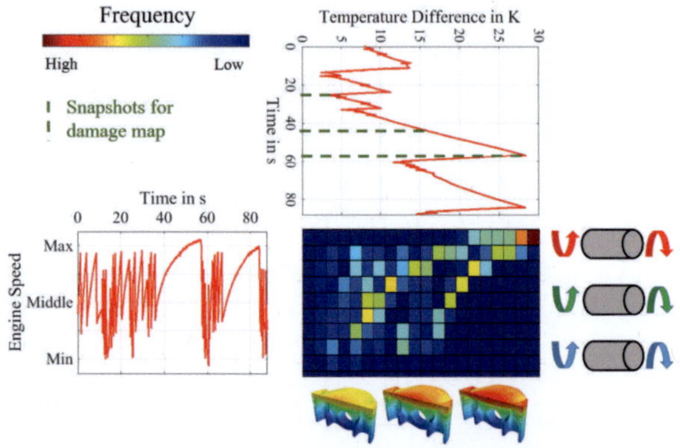

Figure 7.2: Definition of meaningful temperature snapshots and engine speed values. Schematic description by means of a histogram for the engine speed under full load condition and the maximum piston temperature.

Therefore, for each engine speed and temperature state, the most damaging stress combination is approximated for 200 consecutive pressure cycles. The corresponding total cycle number without fracture $N_{c,k}$ for the pressure cycle k is determined according to the FEM method in Chapter 7.2.1, and the fatigue model in Chapter 7.3. According to a linear damage accumulation, the damage of a single pressure cycle k, and the corresponding representative damage state $\tilde{D}$, which is used for the damage maps, are defined as

$$D_k := \frac{1}{N_{c,k}}, \tag{7.1}$$

$$\tilde{D} := \frac{1}{K} \sum_{k=1}^{K} D_k. \tag{7.2}$$

According to a statistical analysis, K is the complete number of investigated pressure cycles for a fixed engine state. In this case, K is 200. The scalar-valued safety factor S_f is similarly defined as

$$\frac{1}{K}\sum_{k=1}^{K} D_k\left(S_f\, p_k\left(\alpha_{\text{cr}}\right)\right) \stackrel{!}{=} \frac{1}{N_{\text{c,t}}}. \tag{7.3}$$

According to Eq. (7.3), the safety factor is defined as a kind of scaling factor for each pressure cycle k, so that the target cycle number $N_{\text{c,t}}$ can be reached. To solve the non-linear, implicitly given Eq. (7.3) for the safety factor, the affine invariant Newton solver NLEQ1 is used.

7.2.1 Scattering pressure cycles

Due to the significant non-linearity of damage, 200 scattering, consecutive pressure cycles are measured for each engine speed and engine setting. For fatigue problems, it is common to define load cycles, consisting of upper and lower loads, or, according to Eq. (7.7), amplitudes with maximum loads. By interpolating, or extrapolating, the high-pressure FEM result between two simulated pressure curves, the upper stress $\boldsymbol{\sigma}_{\text{u,ik}}$ for each pressure cycle $p_k\left(\alpha_{\text{cr}}\right)$ is determined. According to Eq. (7.4), this upper stress is approximated for three, discrete crank angles $\alpha_{\text{cr,i}}$ in the combustion stage. The indices lp (low pressure) and hp (high pressure) describe the two aforementioned enveloping pressure curves $p_{\text{lp}}\left(\alpha_{\text{cr}}\right)$ and $p_{\text{hp}}\left(\alpha_{\text{cr}}\right)$. The corresponding stress states, $\boldsymbol{\sigma}_{\text{lp}}$ and $\boldsymbol{\sigma}_{\text{hp}}$, generally depend on the temperature field, the engine speed, and the engine setting, respectively. In case of the lower load, due to negligible scattering effects in the low pressure stage during the gas exchange, it can be assumed that the lower stress state $\boldsymbol{\sigma}_{\text{l,j}}$ does not depend on the cycle k. The index j describes the discrete crank angle $\alpha_{\text{cr,j}}$ in the low pressure stage, namely the acceleration at the top dead center, and the maximum side force from the reverse direction. These loads also depend on the temperature field, the engine speed, and the engine setting. Afterwards, the most damaging combination is individually determined for each pressure cycle $p_k\,(\alpha_{\mathrm{cr}})$. The method, which is based on the Rainflow counting theory (Radaj and Vormwald, 2007), is illustrated in Fig. 7.3. In summary, for each point in the damage map, e.g., for a given temperature field and engine speed, following algorithm is applied:

1. Calculate potential lower stress tensors $\boldsymbol{\sigma}_{\mathrm{l,j}}$ with $j = 1, 2$.
2. For each pressure cycle k: Calculate potential upper stress tensors $\boldsymbol{\sigma}_{\mathrm{u,ik}}$ with $i = 1, 2, 3$ according to:

$$\boldsymbol{\sigma}_{\mathrm{u,ik}} = \boldsymbol{\sigma}_{\mathrm{lp}} + \left(\frac{p_k}{p_{\mathrm{lp}}} - 1\right)\left(\boldsymbol{\sigma}_{\mathrm{hp}} - \boldsymbol{\sigma}_{\mathrm{lp}}\right)\left(\frac{p_{\mathrm{hp}}}{p_{\mathrm{lp}}} - 1\right)^{-1}, \tag{7.4}$$

$$\boldsymbol{\sigma}_{\mathrm{lp}},\ \boldsymbol{\sigma}_{\mathrm{hp}},\ p_k,\ p_{\mathrm{lp}},\ p_{\mathrm{hp}} = f\left(\alpha_{\mathrm{cr,i}}\right).$$

   Remember that the two enveloping pressure curves p_{lp} and p_{hp}, as well as the corresponding FEM results $\boldsymbol{\sigma}_{\mathrm{lp}}$ and $\boldsymbol{\sigma}_{\mathrm{hp}}$, depend on the engine configuration, e.g., the temperature field and the engine speed.
3. Calculate for each pressure cycle k the most damaging stress combination, e.g., solve following minimization problem:

$$N_{\mathrm{c},k} = \min_{i,j} N_{\mathrm{c}}(i, j), \tag{7.5}$$

   with the total cycle number without fracture N_{c} according to Eq. (7.7).

Again, to solve the non-linear, implicitly given Eq. (7.7) for the total cycle number $N_{\mathrm{c},k}$, the open-source affine invariant Newton solver NLEQ1 is used within the own Matlab code. As can be seen later, due to possible changes of the critical combination of upper and lower stresses, not differentiable kinks in the solution space can occur.

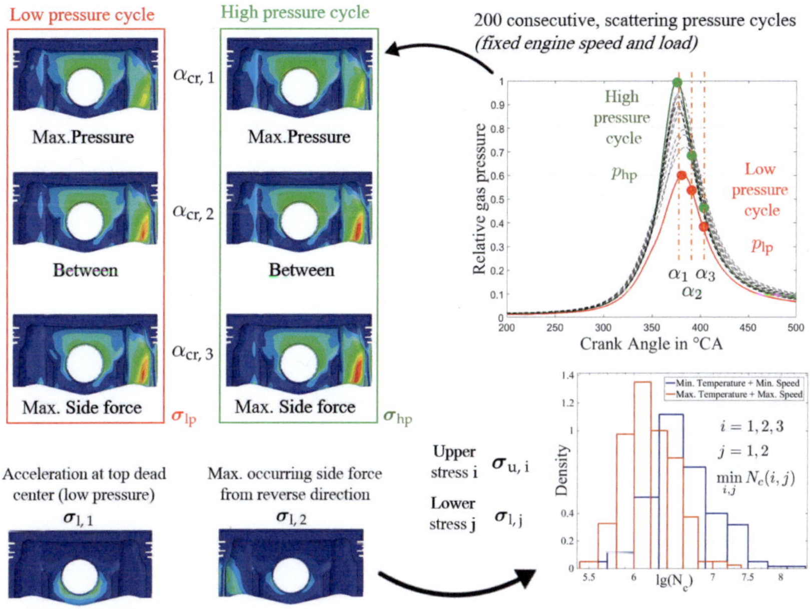

Figure 7.3: Model reduction by interpolating FEM results between two pressure cycles according to Eq. (7.4). For each pressure cycle p_k (α_{cr}), the most damaging combination of low (j) and high (i) pressure load is determined. As an example, in the lower right-hand corner, two damage histograms are shown for different temperature fields and engine speeds.

7.2.2 Initial plasticity simulation

In the present study, it has been noticed that, due to temporary pressure overloads, e.g., knocking cycles, some local areas of the piston undergoes plastic deformations. In particular, this applies to the areas of the piston-pin assembly, as well as the piston skirt bridge. As reported in Chen et al. (2017), these residual stresses have an influence on the mean stress, and, therefore, on the fatigue behavior. To estimate these additional mean stresses, an initial, elastoplastic simulation with a maximum occurring peak pressure and temperature state, as well as a minimum engine speed, is performed. This worst-case scenario should represent the maximum, yet realistic, possible plasticizing. To do this, the maximum integral force on the piston crown, as well as the the corresponding maximum side force, is investigated. For the FEM simulation, the associated flow rule according to von Mises is used as a simple yield criterion with a temperature-dependent yield stress.

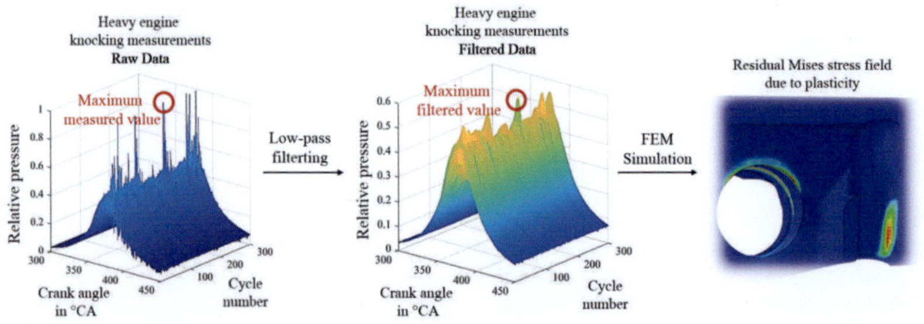

Figure 7.4: Filtering method for the investigation of plastic residual strains and stresses due to recurrent knocking cycles. In the lower right-hand corner, a qualitative contour plot of the residual Mises stress is shown.

For typical piston temperatures and monotonous strain, hardening effects are weakly pronounced for the used material Al2618-T6 (Khalil, 2014). Aging effects are neglected. Fig. 7.4 shows the method to determine reasonable boundary conditions. To achieve heavy knock conditions, a single cylinder engine is run with high intake air temperatures, a low engine speed, and large ignition angles. In sum, over 2000 pressure cycles are investigated with regard to knocking. To approximate the integral force on the piston crown under knocking conditions, the eigenfrequencies of the combustion chamber are used as a low-pass filter (Hettinger and Kulzer, 2009). Afterwards, the maximum observable, but filtered, peak pressure is selected as a homogenous stress boundary condition. In case of the investigated engine, as can be seen in Fig. 7.4, the maximum measured value of the pressure transducer has to be multiplied by a factor of 0.6. This approach is only reasonable for points which are mainly loaded by the resulting forces. The situation is completely different for areas which are directly in contact with the pressure waves, e.g., the piston ring grooves. In this case, increased heat transfer and local pressure oscillations result in higher plasticity and damage (Ferraro et al., 1996). In addition, the oil tightness gets lost, resulting in a self-strengthening effect. For these points, the resulting shift of the mean stresses is not considered.

7.3 Fatigue modeling

7.3.1 Modeling of crack initiation

To model accurately the underlying fatigue mechanisms, a distinction between crack initiation and propagation is reasonable. As an example, one can find some extensive observations on crack initiation for the aluminium alloy 2219-T851 in Morris et al. (1976). For many wrought aluminium alloys, independent from the vague definition of crack initiation, especially in polycrystalline metals (Jiang and Feng, 2004), it can be shown that it is the dominant fatigue regime during high cycle fatigue. A good review abut the physics of crack initiation is given in Sangid (2013). Modeling the physically short and long crack behavior, by the example of the two aluminium alloys 2024-T3 and 7075-T6 under fully reversed, alternating loading, Santus and Taylor (2009) has determined the transition between the two fatigue regimes at approximately 10^4 cycles. To do this, in order to get the initiaion time, the physically short, and long, crack propagation cycles are subtracted from the entire fatigue life. Due to this hint, which is also supported by other references, e.g., (Tanaka and Mura, 1982b) and (Zhao and Jiang, 2008), and due to its practical applicability, the fatigue model according to Papadopoulos (Papadopoulos, 1996) is selected, and, according to Section 7.3.2, additionally modified for own needs. As an example for high cycle fatigue conditions, reasonable results could be achieved for a similar aluminium alloy 7075 in the T651 state (Wang et al., 2014). A further excellent modeling approach for crack initiation with the help of dislocation pile-ups can be found in Tanaka and Mura (1981). A more detailed modeling of various crack initiation mechanisms at inclusions are described in Tanaka and Mura (1982b): slip-band crack emanating from a debonded inclusion, the inclusion cracking due to impinging of slip bands, and the slip-band crack emanating from an uncracked inclusion. Analogously, several semi-empirical formulae for notch and crack-like notch fatigue problems can be described with the dislocation dipole accumulation model in Tanaka and Mura (1982a).The model according to Papadopoulos uses an average measure of the accumulated plastic strain within a representative macroscopic volume element.

The strain localisation is described by the Lin-Taylor approach. In addition, based on the approximately isotropic elastic behavior of an aluminium single crystal, the stiffness tensor on the microscopic scale is assumed to be the same as the effective, macroscopic stiffness. Crystal plasticity itself is described with the help of Schmids law, and the texture is required to be isotropic. The original model takes isotropic and kinematical hardening effects into account, and uses the following root mean square as a crack initiation criterion:

$$\gamma_{\mathrm{p,c}} = \sqrt{5}\sqrt{\frac{1}{8\pi^2}\iiint \gamma_{\mathrm{p,ac}}^2(\phi,\theta,\xi_s)\,\mathrm{d}\phi\,\sin(\theta)\mathrm{d}\theta\,\mathrm{d}\xi_{\mathrm{s}}}, \tag{7.6}$$

with an averaged, critical plastic strain $\gamma_{\mathrm{p,c}}$, which can be a temperature-dependent material parameter. In addition, according to experimental observations, it should depend on the average normal stress of the cutting planes. It can be shown that this is exactly the hydrostatic stress of the effective, macroscopic stress $\mathrm{tr}(\boldsymbol{\sigma})/3$. From a physical point of view, it represents the complex process of the formation of free surfaces. In addition, according to the vague differentiation between crack initiation and propagation, it considers the effect of very small embryo-cracks. In general, with this additional term, the mean stress effects on metal fatigue is considered: a good review about this topic can be found in Dowling (2004). As another example, Desmorat et al. (2015) added an additional Drucker-Prager criterion into the plastic potential on the microscale. The accumulated plastic shear strain on each cutting plane and slip direction $\gamma_{\mathrm{p,ac}}$ is described by the spherical coordinates ϕ and θ, as well as the slip direction angle ξ_s. Neglecting the initial, micro yield stress, one can reduce Eq. (7.6) to the final form:

$$\beta_{\mathrm{l}} - \alpha_{\mathrm{l}}\mathrm{tr}(\boldsymbol{\sigma})|_{\mathrm{max}} = \sqrt{5}\sqrt{\frac{1}{8\pi^2}\iiint \|\langle\boldsymbol{\tau}_{\mathrm{a}}\rangle\|^2(\phi,\theta,\xi_s)\,\mathrm{d}\phi\,\sin(\theta)\mathrm{d}\theta\,\mathrm{d}\xi_{\mathrm{s}}}, \quad (7.7)$$

$$\mathrm{tr}(\boldsymbol{\sigma})|_{\mathrm{max}} = \max_{t\in\mathrm{C}} \mathrm{tr}(\boldsymbol{\sigma}). \quad (7.8)$$

with the macroscopic, resolved shear stress amplitude $\langle\boldsymbol{\tau}_{\mathrm{a}}\rangle(\phi,\theta,\xi_s)$ on the corresponding cutting plane and slip direction. Its detailed definition for arbitary three-dimensional stress states is based on the stress projection onto the gliding direction, and can be looked up in Papadopoulos (1996). In Eq. (7.8), the maximum value of the hydrostatic stress during one cycle C, $\mathrm{tr}(\boldsymbol{\sigma})|_{\mathrm{max}}$, is used. Together with the right-hand side of Eq. (7.8), it can be seen as an equivalent stress, consisting of the stress amplitude and the maximum hydrostatic stress.

At a closer look, Eq. (7.8) describes an energy criterion for crack initiation. The significant dependence is the square of the stress amplitude. Compared with the equation (36) in Tanaka et al. (1981), the above model describes, in principle, a pileup model for irreversible dislocation movements. In case of a douple pileup, the fatigue criterion has the form

$$(\Delta\tau_{\mathrm{D}} - 2k_{\mathrm{f}})^2 \propto \frac{G w_{\mathrm{s}}}{a_{\mathrm{g}}(1-\nu)N_{\mathrm{c}}}. \quad (7.9)$$

Equation (7.9) can be derived by solving the dislocation equilibrium condition with the help of the inverse formula of Muskhelishvili. The specific fracture energy for a unit area is w_s, and the average grain size is a_g. In addition, due to the fact that the dislocation distribution depends on the material properties, the shear modulus G, as well as the Poisson ratio ν, is included. The range of the shear stress during one cycle is described by $\Delta\tau_D$, whereas the friction stress for the dislocation movement is k_f. As can be seen, the Petch-type equation for the grain size dependency of fatigue strength is contained. Changes in the mean stresses could be considered by adjusting the friction stress. Therefore, the second term of the left-hand side of Eq. (7.8) can be interpreted as a change of the friction stress for the dislocation movement.

7.3.2 Model calibration for Al2618-T6

In the context of a Woehler line approach, to fit given temperature-dependent Woehler lines, the material parameter β_l is assumed to depend on the temperature T and the total cycle number without fracture N_c: $\beta_l = \beta_l(T, N_c)$. It is easy to show that β_l is the strength at a fully reversed torsion fatigue test. In particular, the explicit dependency on the last argument implicitly contains possible isotropic strengthing effects. This overcomes, therefore, the difficult determination of microscopic strengthing coefficients. In addition, due to the observable non-linear effect of mean stresses on the fatigue life, α_l is assumed to be stress-dependent: α_l=$\alpha_l(\mathrm{tr}(\boldsymbol{\sigma})|_{\max})$. For ductile materials, this mean stress effect increases disproportionately with higher mean stresses (Pallarés-Santasmartas et al., 2018).

For calibration purposes, temperature-dependent alternating and pulsating tensile test results, as well as bending test results, can be found in: Lumm (1966), Robinson et al. (2003), Kaufman (2008) and Doyle (1964). Separating the influence of temperature and total cycle number, following initial function for the parameter β_l is chosen:

$$\beta_l(T, N_c) = \tilde{\beta}_l(N_c)\tilde{\beta}_T(T, N_c), \tag{7.10}$$

$$\tilde{\beta}_l(N_c) = \beta_a + \beta_b \log_{10}\left(\frac{N_c}{N_{c,r}}\right) + \beta_c \log_{10}^2\left(\frac{N_c}{N_{c,r}}\right) \tag{7.11}$$

$$+ \beta_d \log_{10}^3\left(\frac{N_c}{N_{c,r}}\right), \tag{7.12}$$

$$\tilde{\beta}_T(T, N_c) = 1 + b_T\left(\frac{T - T_r}{T_r}\right) + c_T\left(\frac{T - T_r}{T_r}\right)^2 \tag{7.13}$$

$$+ d_T\left(\frac{T - T_r}{T_r}\right)^3 + e_T \log_{10}\left(\frac{N_c}{N_{c,r}}\right).$$

It has been noted that the temperature influence depends on the load level, and, therefore, the last term in Eq. (7.13) is necessary. The higher the cycle number, the smaller the reduction factor, and, therefore, the coefficient e_T has to be negative. The model constants from β_a to e_T are determined according to a weighted non-linear least square optimization. The weights are given through the inverse of the square of the stresses, and, therefore, a percentage deviation is used as optimization criterion. For dimensional reasons, the subscript r describes reference states. In this study, 10^7 and 293 K are chosen for the cycle number and the temperature. Analogously, in case of an uniaxial stress state, the parameter α_l is modeled according to:

$$\alpha_l(\sigma_u) = \alpha_a + \alpha_b\left(\frac{\sigma_u}{\sigma_{u,r}}\right)^2 + \alpha_c\left(\frac{\sigma_u}{\sigma_{u,r}}\right)^4. \tag{7.14}$$

According to Eq. (7.8), the upper stress for uniaxial loading is given by σ_u. Using the lower stress σ_l, the stress ratio is defined as $R_s = \sigma_l/\sigma_u$. Again, σ_{ref} describes the reference stress and is set to 130 MPa at 293 K. Due to the experimental observation that the slope within the Haigh diagram is correlated with the stress ratio between the fatigue limit and the static tensile strength, the model parameters within the initial function for α_l are assumed not to depend on the temperature. Increasing the temperature from 373 K to 573 K, both quantities reduce nearly to the same extent: 38 % for the 1e7 fatigue limit, and 43 % for the tensile strength. Therefore, for temperature dependencies, according to the decrease of the fatigue strength, only the reference stress has to be scaled. This method is in accordance with Hänel and Wirthgen (1994): suggesting a constant mean stress sensitivity for higher temperatures, only the bounds of the Haigh diagram have to be modified.

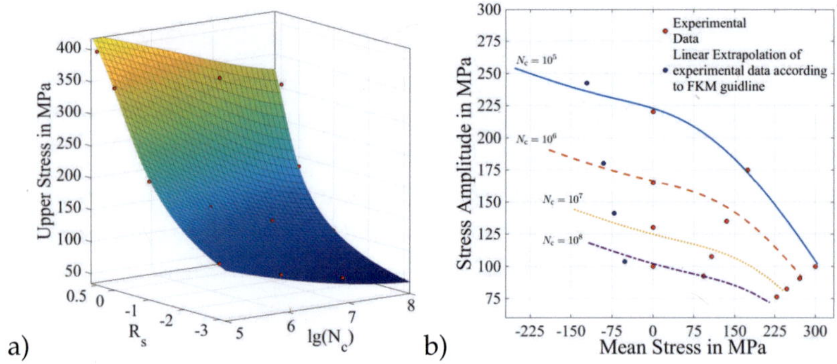

Figure 7.5: Calibration results for a temperature of 293 K. a) Calculated upper stresses. b) Based on the polynomial approach (7.14), calculated Haigh diagram for Al2618-T6 for stress ratios of $-\infty \leq R_s \leq 0.5$.

For ductile materials, there exists a non-linear mean stress effect. Equation (7.14) ensures a quite linear mean stress effect for low stress levels, as well as a higher sensitivity with increasing mean stresses. From a physical point of view, it has to be ensured that the parameter α_l is positive. In addition, the complete term $\alpha_l \mathrm{tr}(\boldsymbol{\sigma})|_{\max}$ should be a continuously increasing function. Together with the calculated upper stresses for different stress ratios and cycle numbers, the predicted Haigh diagram can be seen in Fig. 7.5. According to the FKM guidline

(Forschungskuratorium and Maschinenbau, 2005), stress values for negative mean stresses are linearly extrapolated within the range of $-\infty \leq R_s \leq -1$. Concerning measured upper stresses, the weighted root mean square of the error, resulting from the optimization process, is 4.5 %. Without losing the linear behavior for small mean stresses, the clear non-linear mean stress effect for stress ratios larger than zero are modeled well with Eq. (7.14). The predicted upper stresses are calculated with Eq. (7.7), (7.11) and (7.14). Therefore, using the fixed point method, the resulting implicit equations have to be solved numerically. Analogously, Fig. 7.6 shows the calibration results for the temperature correction.

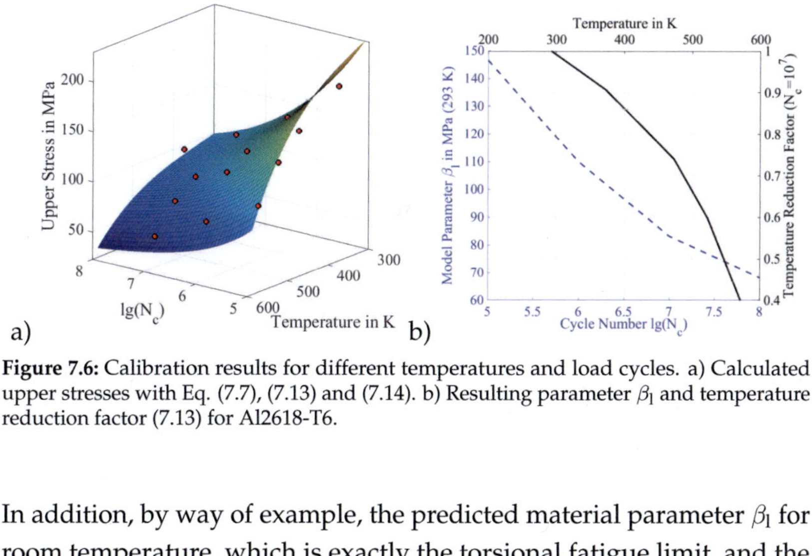

Figure 7.6: Calibration results for different temperatures and load cycles. a) Calculated upper stresses with Eq. (7.7), (7.13) and (7.14). b) Resulting parameter β_1 and temperature reduction factor (7.13) for Al2618-T6.

In addition, by way of example, the predicted material parameter β_1 for room temperature, which is exactly the torsional fatigue limit, and the corresponding temperature reduction factor for a cycle number of 10^7 is shown. In this case, concerning measured upper stresses, the weighted root mean square of the error, resulting from the optimization process, is about 8 %. As a kind of validation for the suggested calibration process, the ratios between measured, uniaxial stress limits and predicted, torsional fatigue limits can be compared with similar aluminium alloys. By way of example, a value of 0.64 for Al7075-T651 can be found in Wang et al. (2014). In Koutiri et al. (2013), 0.7 is reported for the aluminum cast alloy A356 in the T6 state. The predicted values are between 0.64 and 0.68 for cycle numbers from 10^5 to 10^7. According to Wang et al. (2014), the model by Papadopoulos is suitable for materials with ratios between 0.577 and 0.8. The resulting fitting Parameter for Al2618-T6 are shown in Table 7.1.

Parameter	**Value**
β_a	83.17
β_b	-21.16
β_c	5.94
β_d	0.34
b_T	-0.41
c_T	0.34
d_T	-0.60
e_T	-0.043
α_a	0.30
α_b	-0.044
α_c	0.0082

Table 7.1: Fatigue fitting parameters for Al2618-T6.

It is important to note that the above presented initial functions and model constants are only valid for a closed parameter space. In this study, the calibration results are physical plausible for cycle numbers between 10^5 and 10^9. At room temperature, approximating the Woehler line with following exponential function

$$\left(\frac{\sigma_\mathrm{u}}{\sigma_\mathrm{u,r}}\right)^{-k_\mathrm{W}} = \left(\frac{N_\mathrm{c}}{N_\mathrm{c,r}}\right), \tag{7.15}$$

the presented calibration method gives values for the inverse slope k_W in the double-log depiction in the range of seven for cycle numbers of 10^6. This is in accordance with European norms, Eurocode (1999). Again, the subscript r defines reference values. Rapidly rising for cycle numbers larger than 10^7, this value increases up to 13.7 at 10^8 stress cycles. This corresponds to the modified Miner rule by which the slope is fictitiously extended by $2k_W - 1$, Haibach (2006). Having a value of 27 at 10^9 cycle numbers, this point is defined as the cut-off limit for which stress cycles below induce no damage. In addition, the available material data ensure the model validity for stress ratios between $-\infty$ and 0.5, as well as for a temperature range between 293 and 573 K. However, because of other damage mechanisms like low cycle fatigue with large plastic macro strains and dominating creep effects, cycle numbers lower than 10^5, and temperatures larger than 573 K, should rather be critically assessed.

7.4 Results and discussion

7.4.1 The effect of different constellations of transient thermal and mechanical loads

Firstly, various damage maps, with the engine speed corners of 6250 and 7250 rpm, are discussed in detail. Therefore, characteristic, vulnerable points at engine pistons are chosen. With a more aggressive ignition angle of 4 °CA, the engine setting is kept constant throughout this chapter. In Fig. 7.7, the damage map is shown for a critical point within the third ring groove on the piston skirt side. The point is located at the bottom radius. For the investigated engine, equivalent to cylinder four in Fig. 5.15, the air fuel ratio of cylinder two is decreased with higher engine speed, resulting, in a statistical sense, in a higher peak pressure and a lower standard deviation. In Fig. 7.7, the pressure histograms for the minimum and maximum engine speed are shown. In this case, for the minimum speed, the mean value is about 15 % lower, whereas the standard deviation is increased by 23 %. However, due to the fact that the acceleration near the top dead center counteracts the in-cylinder pressure, the representative damage decreases non-linearly with increasing engine speed. Remember that the acceleration is in quadratic proportion to the engine speed. This effect can also be seen in the pressure Woehlerline: with higher engine speeds, for a constant peak pressure, the calculated cycle numbers N_c are shifted to the right.

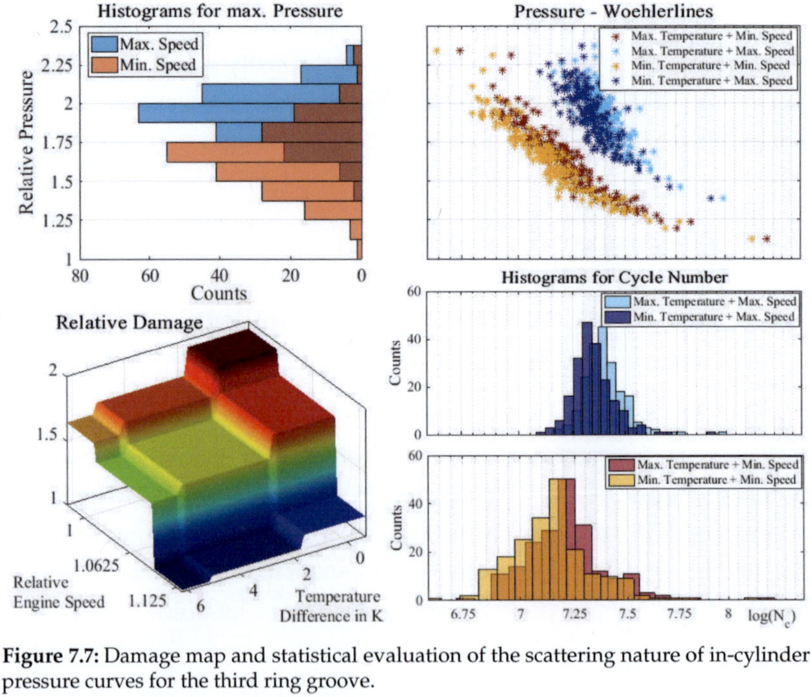

Figure 7.7: Damage map and statistical evaluation of the scattering nature of in-cylinder pressure curves for the third ring groove.

Based on the interpolation method from Eq. (7.4), each dot represents a pressure cycle. However, for a given temperature field and engine speed, the peak pressure is only of limited value: the scattering nature on a horizontal line, e.g., at a constant maximum pressure, results from different pressure curves, and, in addition, various combinations of upper and lower stress states. The last effect is visualized in Fig. 7.8. Firstly, in a statistical sense, with lower peak pressure values, the critical upper stress is shifted towards later crank angles, e.g., towards the maximum side force state. This is reasonable due to the fact that lower peak pressure values often result in a higher pressure level for later crank angles: the combustion process is delayed. Secondly, with lower engine speed, the critical, lower stress is shifted from the maximum acceleration state at gas exchange towards the maximum side force state. Of course, the acceleration force decreases non-linearly with reduced engine speed.

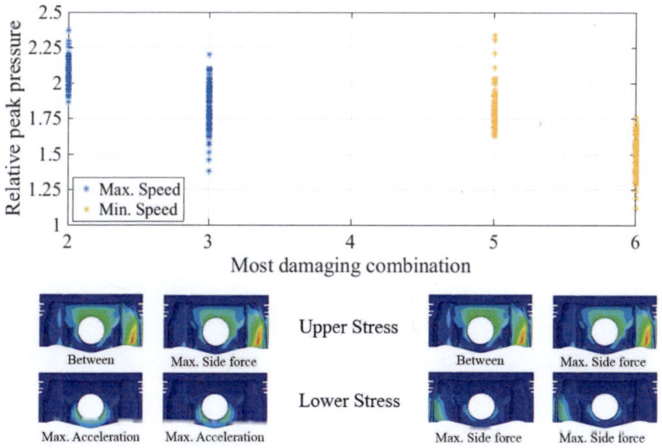

Figure 7.8: According to Fig. 7.3, the dependency of the most damaging combination between upper and lower stress for the critical point within the third ring groove. Each dot represents a pressure cycle.

Concerning the temperature dependency, it can be observed that the representative damage state is decreased with higher temperatures. By the example of the third ring groove, the temperature difference, between a coasting state at curves and a full load situation at the end of straight, is not very high. With approximately six Kelvin, it is a single digit change. Therefore, the temperature reduction factor in Eq. (7.13) is of less importance. By the example of pistons, in relation to the local stress states, temperature fields with higher maximum values near the combustion chamber can increase or decrease the mechanical loads. In this example, at higher temperatures, the mean stress is reduced by approximately 20 MPa. According to the equations (7.7) and (7.8), each three-dimensional load combination can be reduced to an equivalent, one dimensional load case. Concerning the pressure Woehlerline, this effect leads to a shift towards larger cycle numbers. For the four corner points of the damage map, the corresponding lifetime histograms can be seen below. In this case, for low engine speeds, the lifetime can differ by one order of magnitude. According to damage accumulation, only a small number of pressure cycles contribute significantly to the overall damage.

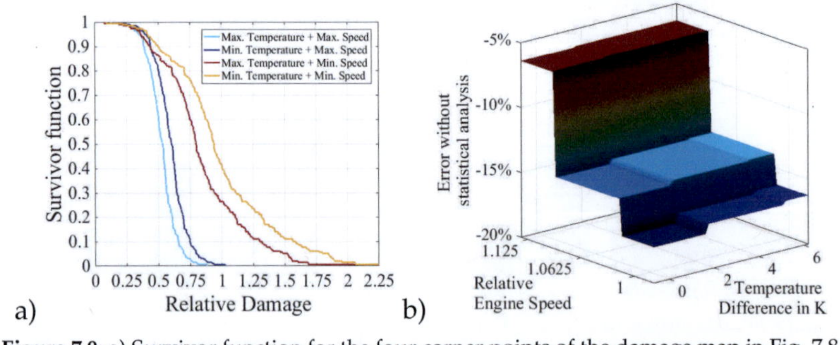

Figure 7.9: a) Survivor function for the four corner points of the damage map in Fig. 7.8. The maximum mean damage value is used as a reference. b) Error without a statistical analysis.

Looking at the damage survivor function $1 - F$ in Fig. 7.9 a), with F as the cumulative distribution function of the damage $D = 1/N_c$, one can notice a large width for low engine speeds. For this case, the probability that a pressure cycle damages the piston more than the maximum mean damage in Fig. 7.8 is about 30 % for the maximum temperature, and about 40 % for the low temperature, respectively. The corresponding error without a statistical analysis is shown in Fig. 7.9 b). For this case, instead of 200 pressure cycles, a fatigue investigation is done for one averaged pressure cycle. Due to the strong scattering character for lower engine speeds, and the non-linearity of the underlying, physical mechanisms, the maximum underestimation of damage is approximately 18 %. In case of higher engine speeds, this error is reduced to 6 %. The temperature influence is negligible. Another relevant area for piston development is the first ring groove. In Fig. 7.10, a damage map with a statistical evaluation is shown for a critical point on the skirt side.

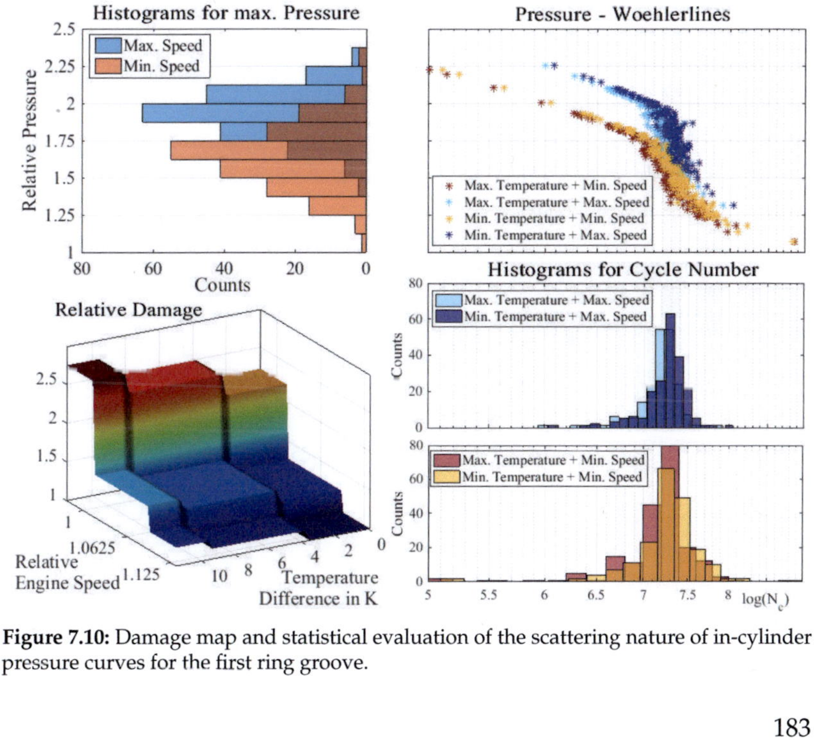

Figure 7.10: Damage map and statistical evaluation of the scattering nature of in-cylinder pressure curves for the first ring groove.

The point is also located at the bottom radius. Compared to the third ring groove, one can notice two significant differences. Firstly, due to the fact that the first ring groove is closer to the combustion chamber, the temperature difference between coasting in curves and firing at the end of straight is larger. In this case, the difference is 10 Kelvin, and, therefore, the temperature reduction factor in Eq. (7.13) has a greater importance. In contrast to the third ring groove, this is the reason why the damage increases with larger temperatures. Compared with a factor of approximately 2 for the most and less damaging constellations, in this case, the factor is 2.8. The second large difference is the strong scattering behavior of the calculated cycle numbers and damage values. By the example of the low engine speed, the lifetime differ by four orders of magnitude. This marked difference is the result of the small gradient in the pressure Woehlerline for high-pressure loads. In addition, for the lowest engine speed, this causes the greater jump in the damage map. In general, one can notice two different behaviors with unequal gradients.

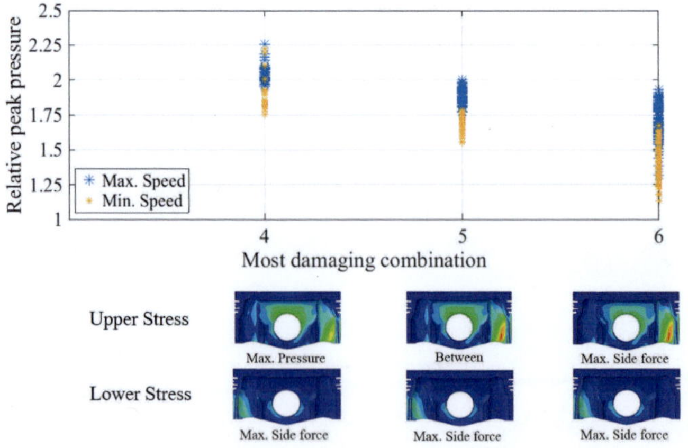

Figure 7.11: According to Fig. 7.3, the dependency of the most damaging combination between upper and lower stress for the critical point within the first ring groove. Each dot represents a pressure cycle.

According to Fig. 7.11, this can be clearly explained by the presence of three relevant damage combinations. For high-pressure values, the small gradient section in the Woehlerline pertains to the combination of maximum pressure and side force. For these crank angles, according to Fig. 7.3, the ratio between various pressure cycle is very pronounced. This also explains the strong scattering behavior of the calculated lifetime. For lower peak pressures, in a statistical sense, the critical, upper stress is more dominated by the side force. Looking at the corresponding survivor function in Fig. 7.12 a), this strong scattering behavior causes a large width. In this case, a logarithmic scale is used for the x-axis. Only a small portion of the pressure cycles is responsible for the representative damage. For low engine speeds, using an averaged pressure cycle, one can, therefore, observe an underestimation of this representative damage in the order of 60 %. The corresponding error map is shown in Fig. 7.12 b). For higher speeds, the error is not that serious. Again, the temperature dependency is negligible.

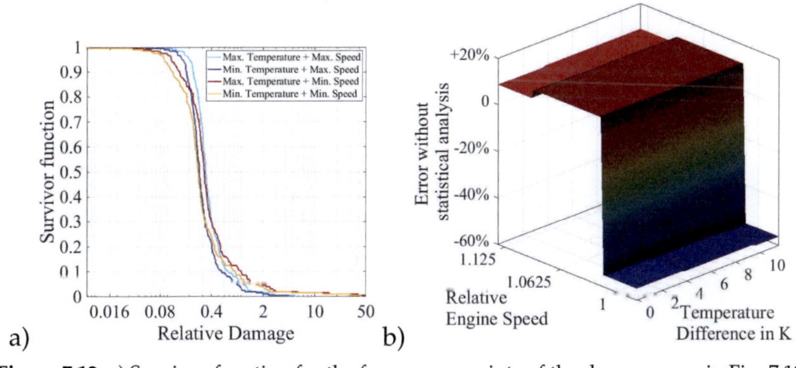

Figure 7.12: a) Survivor function for the four corner points of the damage map in Fig. 7.10. The maximum mean damage value is used as a reference. b) Error without a statistical analysis.

The next characteristic point is the piston hub. According to Fig. 7.4, this area is affected by a plastic deformation under knocking conditions. The investigated point is inclined sidewards by approximately 45 degrees towards the piston skirt. The damage evaluation is shown in Fig. 7.13. Similarly to the other two points, under transient engine conditions, the damage map range is within a factor of approximately 2.5. As expected, the temperature influence is not very pronounced. In contrast to a larger reduction factor according to Eq. (7.13), with relation to the local stress state, higher temperatures relieve the piston hub. However, the engine speed changes significantly the representative damage. Although the pressure Woehlerlines do not fundamentally differ from each other, higher engine speeds cause significantly larger damage values. Looking at the lifetime histograms, this dependency is mainly caused by higher peak pressure values for higher engine speeds.

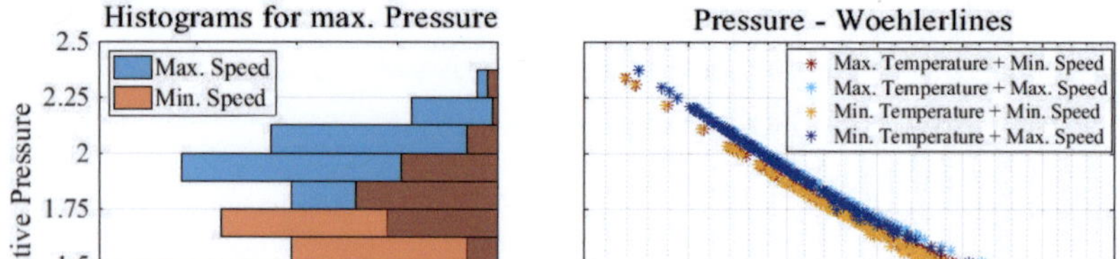

Figure 7.13: Damage map and statistical evaluation of the scattering nature of in-cylinder pressure curves for the piston hub.

Remember that, according to Fig. 5.15, the air-fuel ratio is decreased with higher engine speeds. In this case, the higher acceleration values are irrelevant. The piston hub is loaded by the resulting force which is the integral over the complete piston crown. The most damaging combinations are mainly consisting of peak pressure and acceleration at the gas exchange. For all engine speeds, only for a small number of pressure cycles with low peak pressure values, the side force state, or the state between peak pressure and side force, is dominant. Again, this is the reason for the well pronounced scattering nature. The corresponding survivor function in Fig. 7.14 a) shows a pronounced width, and, due to the above discussed reasons, the curve is mainly affected by the engine speed. The error which is made by using an averaged pressure cycle is always quite large. According to Fig. 7.14 b), this error is approximately between 35 % and 50 %. For a lower engine speed, the wider lifetime distribution in Fig. 7.13 causes a larger error.

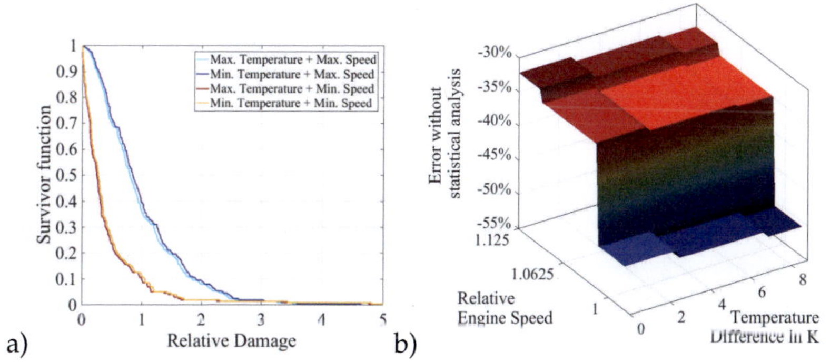

Figure 7.14: a) Survivor function for the four corner points of the damage map in Fig. 7.13. The maximum mean damage value is used as a reference. b) Error without a statistical analysis.

The last investigated point is the critical point on the piston crown, offset from the centre. In this case, the stresses are not that high. Therefore, instead of the representative damage value, the safety factor according to Eq. (7.3) is shown in Fig. 7.15. The target cycle number is 4.2e6. Due to the direct contact with the combustion gas, the temperature fluctuations during a race lap are significant.

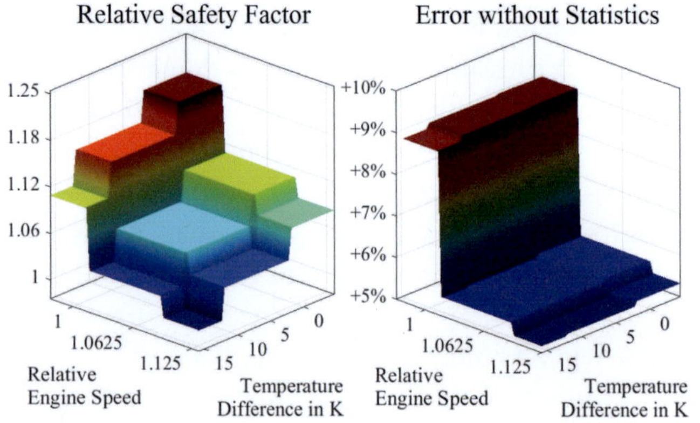

Figure 7.15: Relative safety factor for the piston crown. The lowest value is set as the reference. The error gives the overestimation by using an averaged pressure cycle.

Therefore, the transient temperature fields are just as important as the various mechanical loads. Clearly, compared to the easing of tension with higher temperature, the reduction factor according to Eq. (7.13) dominates. Again, despite of a larger acceleration, higher engine speeds cause more damage due to larger pressure values: the deflection of the complete piston is dominated by the resulting force which is the integral over the complete piston crown. The error by using an averaged pressure cycle is not that high. In general, an overestimation for the safety factor, e.g., an underestimation for the damage, can be observed. For low engine speeds, due to less peak pressure values, and a corresponding larger width in the pressure Woehlerline, this error is a bit increased.

7.4.2 Damage influence of various engine settings

In the following chapter, with regard to the piston lifetime, the influence of various engine settings is investigated. Remember that both, mechanical loads in form of different pressure curves and transient temperature fields, depend on this parameterization. According to Fig. 7.1 and 7.2, safety maps are compared. Again, the target cycle number is 4.2e6. However, for the purposes of time and cost reduction, reduced maps with a limited number of speed and temperature grid points, namely the minimum and maximum values, are used. In addition, according to the exemplary histogram in Fig. 7.2, the combination between the maximum temperature and the minimum engine speed can be omitted.

Variation of the ignition system

Firstly, with regard to the safety factor, the two different ignition systems, namely the spark plug configuration and the jet-ignition, are compared. In the latter case, the difference of the simulated maximum piston temperature at the end of straight is 2 K higher. In Fig. 7.16 a), the corresponding histograms for the maximum in-cylinder pressure are shown for two different engine speeds. With jet-ignition, the peak values are, in a statistical sense, clearly higher, and the histograms are shifted to the right. Especially with a low engine speed, the difference is well pronounced. In addition, the standard deviation is much smaller. To be more precise, in case of the higher engine speed, the mean value is approximately 8 bar larger. Correspondingly, for the lower engine speed, the expectation value is 17 bar higher. For the standard deviations, the differences are 6.5 and 6.3 bar, respectively. In summary, with jet-ignition, the burn rate is faster, and, with regard to peak pressure values, the combustion process is more stable. The corresponding influence on the safety factors can be seen in Fig. 7.16 b) for the aforementioned piston locations. Therefore, the safety factor with jet-ignition is divided through the safety factor with spark plug configuration. Excluding the ring groove one, the jet-ignition causes higher damage values, and the safety factors are smaller.

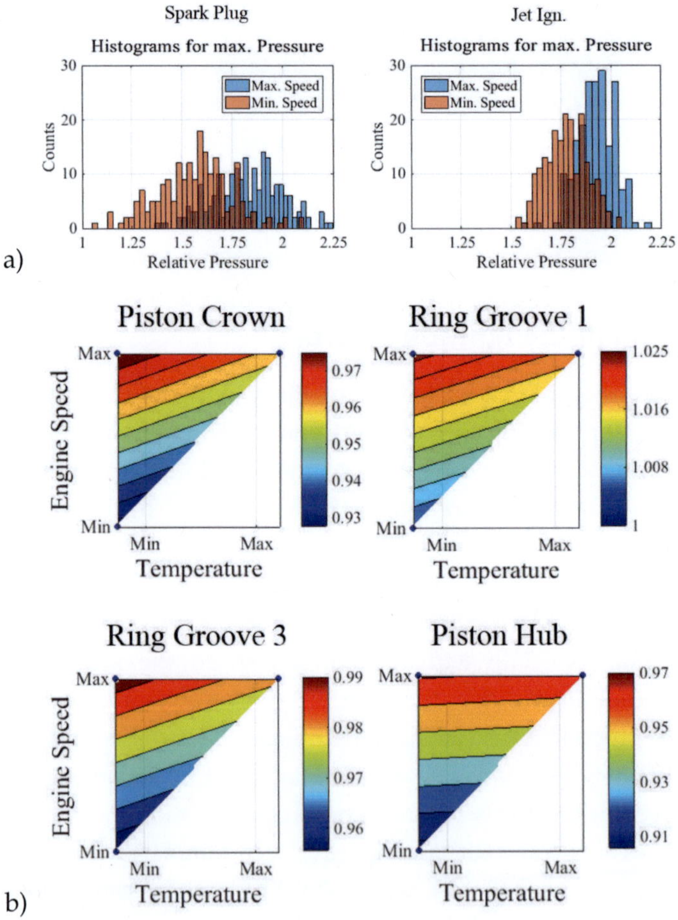

Figure 7.16: Change of ignition system. a) Histograms for the maximum in-cylinder pressure. b) Ratio between the safety factors. Values smaller than one mean less safety under jet-ignition.

With lower engine speeds, and higher temperatures, there exist a higher sensitivity towards the ignition system. Remember that the temperatures are higher with jet-ignition. In addition, according to Fig. 7.16 a), the maximum in-cylinder pressure differs strongly for low engine speeds. The closer the point is located to the combustion chamber, the steeper the safety-isolines are. This is reasonable because of a higher sensitivity towards temperature amplitudes. In case of the lower engine speed, the largest difference of about 10 % can be observed for the piston hub. In this case, the mechanical load is determined by the complete integral force on the piston crown. For this location, together with the third ring groove, the values are between 1 % and 4 %, as well as 3 % and 7 %, respectively. However, for the ring groove one, one can observe an exception. With jet-ignition, the safety factors are larger. According to Fig. 7.10, the strong scattering behavior of the calculated cycle numbers is the reason. As already discussed in Section 7.4.1, this is a special characteristic of the investigated point within the first ring groove. In fig 7.17, the pressure Woehlerlines are compared for the two ignition systems. Despite of the higher mean value of the maximum in-cylinder pressure, the jet-ignition system results in lower damage values.

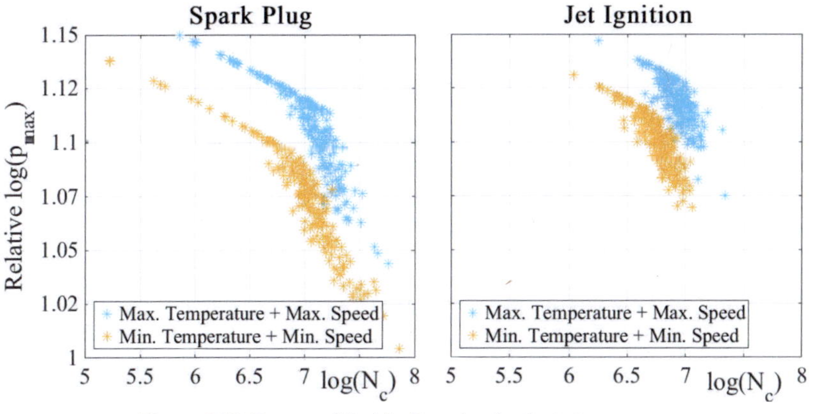

Figure 7.17: Pressure Woehlerlines for the first ring groove.

In case of the spark plug configuration, the standard deviation is approximately twice the value of jet-ignition. The safety factors, together with the calculated histograms for the total cycle numbers, are shown in Fig. 7.18. Theses histograms are calculated for the scaled pressure curves, e.g., with the safety factor. Concerning the safety factors within the engine map, the qualitative progression remains the same for different ignition systems. However, according to Fig. 7.16 a), the scattering behavior of the total cycle numbers vary widely.

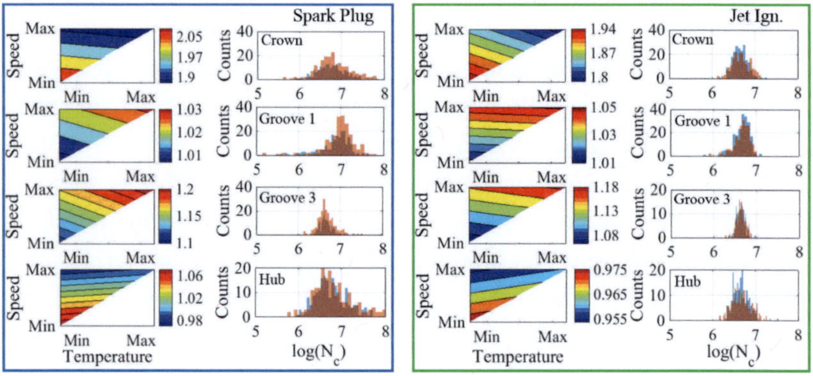

Figure 7.18: Influence of the ignition system on the safety factors of various piston locations. In addition, the histograms for the calculated cycle numbers are given. The color assignment for the two engine speeds is the same as in Fig. 7.16 a).

Lambda variation - jet-ignition

To investigate the effect of air-fuel ratio, the lambda variation under jet-ignition conditions is chosen. Therefore, according to Fig.6.2, the basic engine setting is compared with the richer combustion mode. In this case, the richer combustion results in approximately 6 K higher maximum piston temperatures at the end of straight. In a statistical sense, the in-cylinder peak pressure values are 30 bar higher with the lower engine speed, and 12 bar for the higher one. In each case, under richer conditions, the standard deviation is approximately one bar lower. The pressure histograms are shown in Fig. 7.19 a). In case of the richer combustion, one important observation are the slightly lower peak values with a higher engine speed.

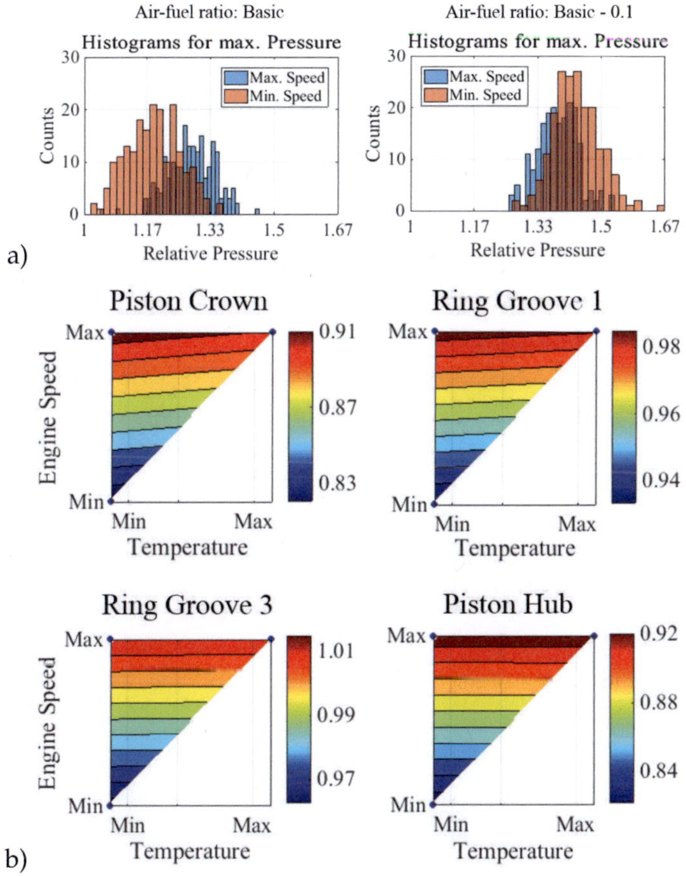

Figure 7.19: Lambda variation under jet-ignition conditions. a) Histograms for the maximum in-cylinder pressure. b) Ratio between the safety factors. Values smaller than one mean less safety under a richer combustion.

The reason for this are the already mentioned cylinder individual air-fuel ratios. According to Fig. 6.3 and 5.15, in case of cylinder two and four, an overall downshift of 0.1 in the air-fuel ratio results in sub-stoichiometric conditions with higher engine speeds: at 7250 rpm, the lambda value is 0.95, whereas it is one for 6250 rpm. For the basic engine configuration, it is always larger than one. In addition, it has to be noticed that the difference in the air-fuel ratio is only 0.75 for the minimum engine speed. The reason is the aforementioned higher knocking tendency and the corresponding self-protection. According to Fig. 7.19 b), for all piston locations, a reduction of the safety factor can be observed. Only in case of the ring groove 3, there is no real difference with a high engine speed. The generally small influence is comparable to the variation of the ignition system. Despite of the aforementioned change towards a sub-stoichiometric combustion, the qualitative progression is comparable to Fig. 7.16 b). However, in all cases, the temperature sensitivity is much smaller. Although the changes are not that large in case of the ring groove one, the safety factor ratios are always smaller than one: the strong influence of the scattering behavior, as can be seen in Fig. 7.17 for the spark plug, is not present for jet-ignition. Again, the highest reduction of the safety factor can be observed for the piston crown and the piston hub. For lambda variations, this reduction is quite large: for low engine speeds, the safety is reduced by almost 20 %. However, in this case, the piston crown is the most sensitive location, and not the piston hub. According to Fig. 7.15, this could be explained by the greater influence of the temperature: lambda variations result in high temperature changes. The safety factors, together with the calculated histograms for the total cycle numbers, are shown in Fig. 7.20. Again, theses histograms are calculated for the scaled pressure curves, e.g., with the safety factor. Compared to Fig. 7.18, the scattering behavior does not differ significantly between different air-fuel ratios. As expected, in case of the piston crown and the piston hub, there exists an opposite effect for engine speed variations. Due to the already mentioned cylinder-individual air-fuel ratio and the corresponding pressure histograms in Fig. 7.19 a), higher engine speeds result in larger safety factors.

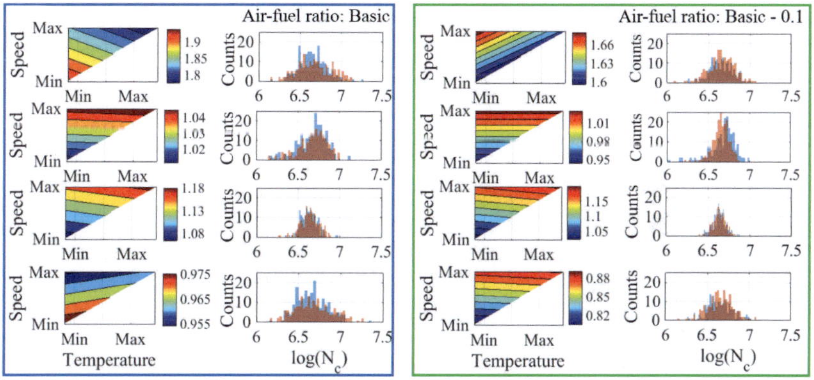

Figure 7.20: Influence of the air-fuel ratio on the safety factors of various piston locations. In addition, the histograms for the calculated cycle numbers are given. The color assignment for the two engine speeds is the same as in Fig. 7.19 a).

Ignition time variation - jet-ignition

To investigate the effect of ignition time, the variation under jet-ignition conditions is chosen. Therefore, according to Fig.6.2, the basic engine setting is compared with an increased ignition angle of 4 °CA. At the end of straight, this earlier ignition time results in a temperature increase of five Kelvin. The mean value of the maximum in-cylinder pressure values differs about 35 bar for the high engine speed. For the lower one, the difference is only 24 bar. Concerning the standard deviation, the basic engine configuration shows an increased value of 1.2 bar for the higher engine speed. At 6250 rpm, the basic setting results in a negligible lower value of about 0.2 bar. The pressure histograms can be seen in Fig. 7.21 a).

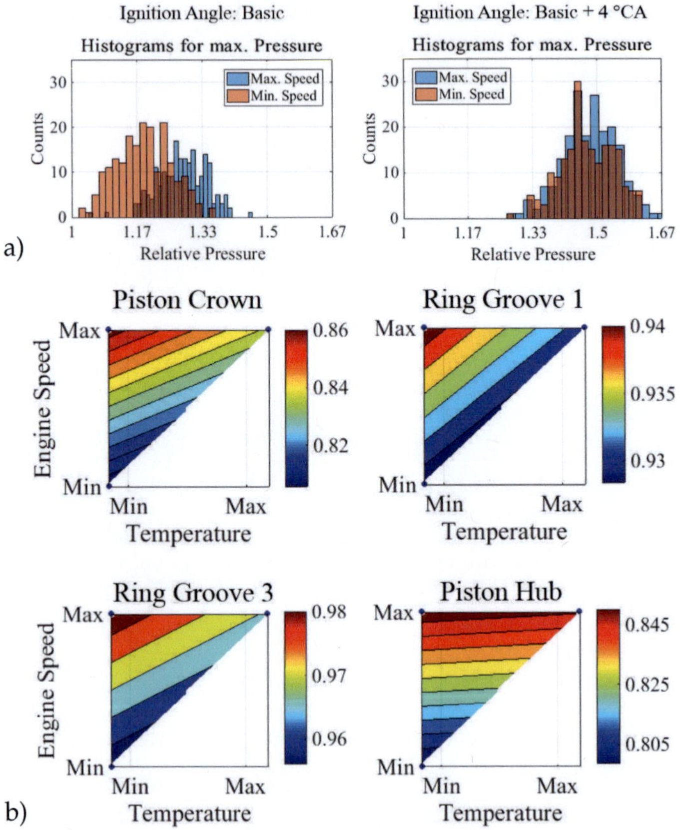

Figure 7.21: Ignition time variation under jet-ignition conditions. a) Histograms for the maximum in-cylinder pressure. b) Ratio between the safety factors. Values smaller than one mean less safety under an increased ignition angle.

Of course, an earlier ignition angle results in significantly larger maximum pressure values. However, it is interesting to notice that there exists no relevant speed dependency for the increased ignition angle. Compared with the air-fuel variation in Fig. 7.19 a), the pressure increase is even larger. At the same time, the temperature increase is comparable. Therefore, as can be seen in Fig. 7.21 b), the safety factors reduce even more. For the piston crown and the piston hub, a reduction of about 20 % can be observed for the lower engine speed. Even for a higher speed, the reduction is approximately 15 % for both locations. Again, both ring grooves react less sensitively. Compared with the air-fuel variation, the safety factor ratios are a little bit lower. The difference is between one and four percentage points. However, the temperature dependency is more pronounced. The safety factors, together with the calculated histograms for the total cycle numbers, are shown in Fig. 7.22. Again, theses histograms are calculated for the scaled pressure curves, e.g., with the safety factor. A little less variation can be seen for the increased ignition angle. However, in both cases, the already noticed increased scattering behavior for the ring groove one can be observed. Here again, only a few pressure cycles determine the lifetime. As expected, in case of the piston crown and the piston hub, there exists an opposite effect for engine speed variations. Due to the very similar pressure histograms in Fig. 7.21 a), the larger acceleration for higher engine speeds has a positive effect on the safety factor.

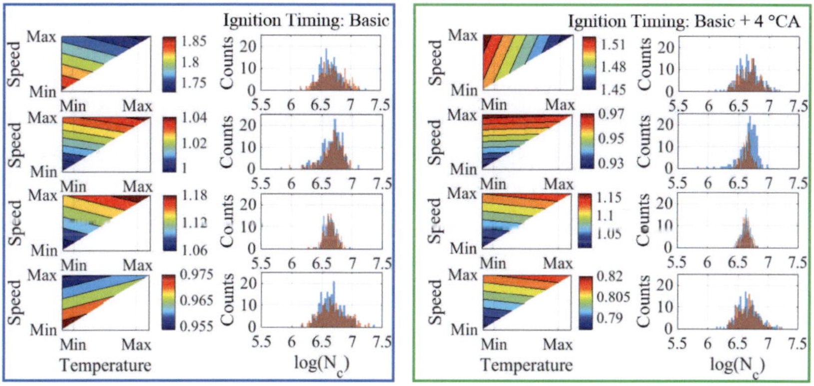

Figure 7.22: Influence of the ignition time on the safety factors of various piston locations. In addition, the histograms for the calculated cycle numbers are given. The color assignment for the two engine speeds is the same as in Fig. 7.21 a).

To investigate possible non-linearities, three different ignition angles with the spark plug are evaluated. The results are shown in Fig. 7.23.

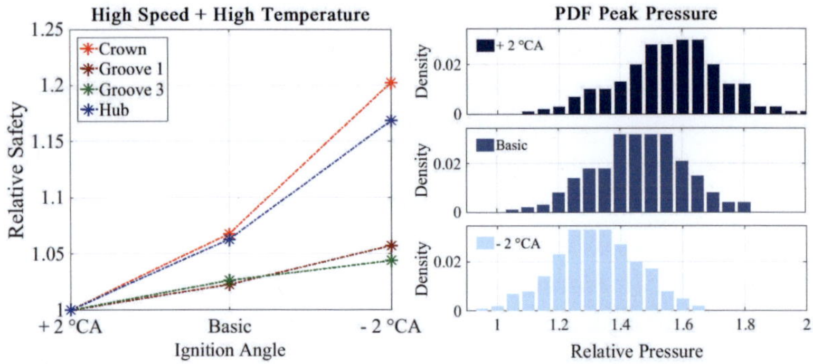

Figure 7.23: Progression of the relative safety factor for three different ignition angles and four different piston locations. In addition, the underlying peak pressure probability density functions are shown. The results are based on a spark plug configuration.

For clarity, only the constellation of maximum engine speed and temperature is shown. In this case, between the lowest and the highest ignition angle, the maximum piston temperature increases by 7.5 Kelvin at the end of straight. For the base ignition angle, the difference is 4.4 Kelvin. Analogously, the average value of the peak pressure is 24 bar higher. Again, for the basic configuration, it is increased by 14 bar. As can be seen in Fig. 7.23, with higher ignition angles, the standard deviation also increases by 1.2 and 4.8 bar, respectively. For the piston crown and piston hub, one can observe a progressive trend of the relative safety factor with decreasing ignition angles. In accordance with Fig. 7.21, these two points are the most sensitive ones. For the first ring groove, this non-linear behavior is pronounced extremely weak. In case of the third ring groove, there exists a linear relationship.

Fuel mass flow variation

For the fuel mass flow variation, according to Fig. 6.2, the spark plug configuration is used. In this case, three different factor levels are available, and, therefore, possible non-linearities could be detected. Between the highest and lowest fuel flow rate, the difference of the maximum piston temperature is 5.3 K at the end of straight. In Fig. 7.24 a), the peak pressure histograms are shown.

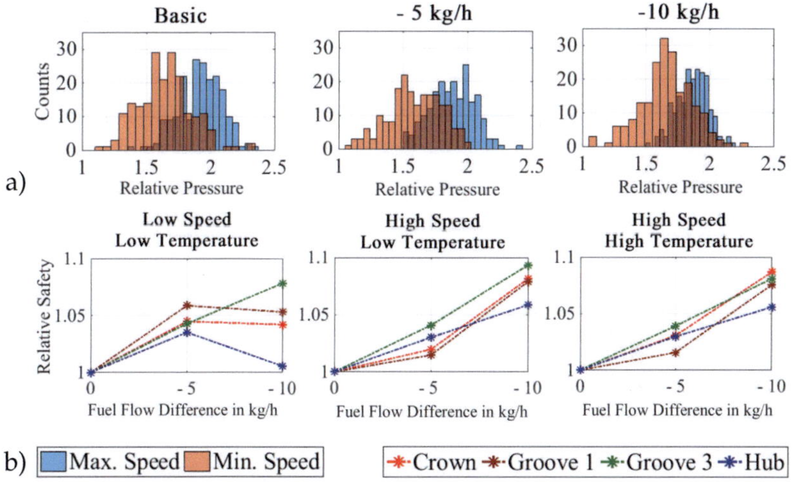

Figure 7.24: a) Peak pressure histograms for three different fuel flow rates and two engine speeds. b) Resulting relative safety factors for four different piston locations. The three graphs correspond to various points within the damage map.

At the higher engine speed, the mean value decreases about 3.4 and 4.6 bar, respectively. At 6250 rpm, the situation is different. Reducing the fuel flow rate about five kilograms per hour, the averaged peak pressure is approximately 3.9 bar lower. However, for a reduction about ten kilograms per hour, the mean value is is 1.5 bar higher than the basic configuration. In accordance with Fig. 6.3 a), the resulting air-fuel ratios for the investigated cylinder are within a tolerance of less than 0.01.

Therefore, this cannot be the reason. However, with a lower fuel flow rate, the ignition angle can be enlarged. Due to its high influence on peak pressure values, e.g., compare Fig. 7.21 a), this could be the reason for the opposite effect at lower engine speeds. Concerning the standard deviations, in case of a reduction about five kilograms per hour, there is no significant change. However, by decreasing the fuel flow rate by ten kilograms per hour at the higher engine speed, the value is 0.7 bar lower. At 6250 rpm, the difference is 3.6 bar. The resulting influence on the safety factor can be seen in Fig. 7.24 b). For each point within the damage map, the multi-stage results are shown for four different piston locations. The largest reduction of the fuel flow is 12.4 %. However, the safety factor never reaches an increase by 10 %. As already explained, in case of the lower engine speed, the rise of the peak pressure at the lowest fuel flow rate causes a lowering of the safety factor. However, the ratio is still larger than one. Concerning non-linearities, one can observe a dependency on the axial piston location: points near the combustion chamber, e.g., the piston crown, show a more progressive course for both higher engine speed points, whereas the hub shows a linear character. This could be explained by the additional influence of the different temperature fields. Near the combustion chamber, the changes are more pronounced. In principle, except for groove three, this different behavior can also be observed for the lower engine speed: in case of the piston hub, compared with the crown, the drop of the safety factor with the minimum fuel flow rate is more pronounced.

Summary of the parameter variation

To summarize the presented design of experiments, safety factor reductions as well as specific reduction values are shown in Fig. 7.25. Therefore, with regard to changes of effective power, a safety-power ratio is defined. In relation to the corresponding power increase, it describes the relative safety reductions in percent. Using the minimum value, the two points in the ring groove one and three are summarized for the sake of clarity.

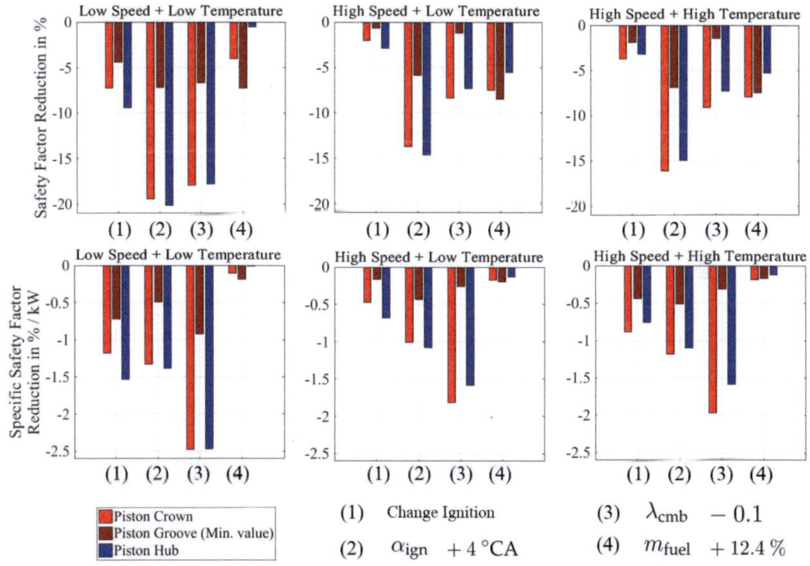

Figure 7.25: Summary of the design of experiments. The upper diagrams show the reduction of individual safety factors for three points within the damage map. In relation to the effective power increase, the corresponding specific values are shown in the lower graph.

Firstly, for most of the variations, one can observe a trend towards higher reduction values at lower engine speeds. With larger engine speeds, the acceleration forces gain in importance, and the engine setting is usually less crucial. In case of the highest fuel flow variation, the already mentioned, enlarged ignition angle causes another progression. However, according to Fig. 7.24, the intermediate level, with an increase of five kilograms per hour, shows again the same behavior. It should also be noted that, in case of the ring grooves, the results suggest a generally lower sensitivity. Within the investigated parameter space, in contrast to the temperature differences in Fig. 6.27, an increase of the ignition angle by 4 °CA causes the largest reduction in the safety factor. This can be explained by the huge increase of the in-cylinder peak pressure values. However, similar to Fig. 6.27, variations of the air-fuel ratio result in the largest differences in the specific values. Nevertheless, concerning the other variations, the gap is not that large anymore. In general, the different behavior of the various parameter variations in Fig. 6.27 b) are similar to those in Fig. 7.25. The influence of the ignition angle and the ignition system is comparable. The extremely low sensitivity to fuel flow variations has also been confirmed by the fatigue calculations.

7.5 Conclusions

Concerning the initial research question, which damage results from different constellations of transient thermal and mechanical loads, and how do various engine settings differ among each other, the result is as follows. All investigated piston locations exhibit a strong dependence on the engine speed. However, for the investigated engine, the increasing in-cylinder gas pressure are partly nullified by the inertial loads. By the example of the piston hub and piston crown, damage is dominated by the gas pressure, whereas the ring grooves react strongly to the inertial loads. With regard to the temperature influence, only the locations near the combustion chamber show a notable dependency. However, although the material resistance against fatigue always falls with increasing temperature, the trend is not clear: different, transient temperature fields can cause various stress states. In case of the piston crown, this temperature effect is comparable to mechanical loads. For both cases, between the two extremes, the safety factor changes are in the range of 10 %. It falls with increasing temperature and in-cylinder gas pressure. Compared to mechanical loads, for the ring groove one and three, the temperature effect is only a fifth and a third, respectively. In contrast to the first, higher temperatures reduce the damage state. In case of the piston

hub, the temperature influence in negligible. In summary, between two corner points, the damage and safety value can differ by a factor of approximately 1.2 and 2, respectively. In addition, due to a possible rearrangement of the most damaging combination between upper and lower stress, the slope of the pressure Woehlerline can significantly change. Moreover, the damage underestimation without a statistical description is evident. Using only one averaged pressure cycle, the maximum error is approximately 60 % for the first ring groove, and 50 % for the piston hub. In case of the third ring groove and the piston crown, the discrepancy is between 17 % for the damage value and 10 % for the safety factor, respectively.
Concerning the change of safety factors for various engine settings, one of the most interesting observations is the fact that, except for the fuel flow variation, the qualitative progression in the engine speed - temperature map is always the same. Concerning the slope of the iso-lines, regardless of the piston location, or the specific comparisons in Fig. 7.16 b), 7.19 b), and 7.21 b), the sign is always the same. In case of fuel flow variations, with specific safety reduction values in the range of 0.2 % per kilowatts, the effect on fatigue is quite small anyway. In general, the differences in the specific reduction values between engine parameters are qualitatively similar to the specific, dynamic temperature differences in Fig. 6.27. For the air-fuel ratio, the largest values, namely at the piston crown and the piston hub, are in the range of 2.5 % per kilowatts at the lower engine speed. For the higher engine speed, they are between 1.5 % and 2 % per kilowatts. Similar to ignition angle variations, a change in the ignition system from spark plug to jet configuration cause changes between 1 % and 1.5 % per kilowatts. Again, with a higher engine speed, the values are in the range of 0.5 % and 1 % per kilowatts. In general, the piston grooves exhibit a smaller dependency on the engine setting. The values are always lower than 1 % per kilowatts.

Chapter 8

Determining water mass flow control strategies for a turbocharged SI engine using a two-stage calculation method[1]

8.1 Outline of the chapter

The research question can be formulated as follows: with regard to engine temperatures near the combustion chamber under highly transient racing conditions, which temperature range, including the temperature deviation, its dynamic behavior and the resulting heat flows, is feasible only by adjusting the water mass flow rate at a constant water inlet temperature? On this basis, what are the differences between a fast lumped capacity model, which is suitable for optimization and sensitivity calculations, and a more compute-intensive, three-dimensional CFD-CHT simulation?

In a first step, the simplified lumped capacity model is developed. With the help of the stationary solution and linearization, a first estimation of suitable strategies and parameters for feed forward and feedback controllers are gained. Afterwards, a Monte Carlo simulation is performed

[1] This chapter is based on the paper "Determining water mass flow control strategies for a turbocharged SI engine using a two-stage calculation method" (Hölz et al., 2019a) **DOI: https://doi.org/10.1016/j.applthermaleng.2018.09.133**

to get optimized control parameters: in a more holistic approach, their influences on engine temperatures, hydraulic flow losses and heat fluxes are investigated. In a second step, the most promising constellations are used for validation and for a more detailed investigation of the aforementioned questions. Therefore, a three-dimensional CFD-CHT engine simulation is used, which is based on the transient method from Chapter 5. In summary, the both, extreme bounds of the full range of model design are compared.

8.2 Optimized thermal management due to various feed forward and control strategies

8.2.1 Method used in this chapter

Various feed forward and control strategies for the engine water mass flow rate are investigated numerically and compared with a mechanical water pump with a fixed transmission ratio.

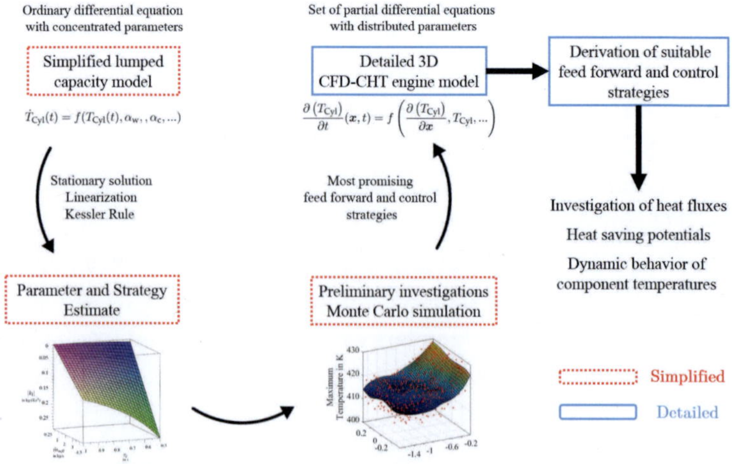

Figure 8.1: Proposed two-stage calculation method for an optimized thermal management.

Fig. 8.1 shows the proposed method. A simplified lumped capacity model of the engine serves as a first parameter and strategy estimation with subsequent, preliminary investigations, including a Monte Carlo simulation. Afterwards, the most promising parameter sets and strategies are investigated with detailed CFD-CHT engine simulations. Special focus is laid on heat saving potentials and the dynamic behavior of diverse component temperatures under a transient drive. Therefore, simplified and the detailed, three-dimensional simulations are carried out for one representative race lap and analysed regarding various criteria such as heat losses, solid temperature fluctuations or the necessary hydraulic power.

8.2.2 Simplified thermal model - lumped capacity model

The used heat transfer models are extracted from Chapter 3 and 5. According to Eq. (5.13), the heat transfer coefficient for the water jacket can be written as:

$$\alpha_{\mathrm{W}}(\boldsymbol{x},t) \approx \alpha_{\mathrm{ref}}(\boldsymbol{x}) \left(\frac{\dot{m}_{\mathrm{W}}(t)}{\dot{m}_{\mathrm{W}}\big|_{\mathrm{ref}}} \right)^{m}. \tag{8.1}$$

The location dependency is omitted because of the lumped capacity model, and, therefore, it serves as a calibration parameter with a certain physical meaning: choosing $\dot{m}_{\mathrm{W}}|_{\mathrm{ref}}$ as a representative mean water mass flow rate, a range of $\alpha_{\mathrm{ref}} \sim 10^4\ \mathrm{W/(m^2K)}$ is used. In this case, the HTC is referred to the water temperature as the reference temperature.
Based on Eq. (5.4), following approach is used for the combustion chamber HTC:

$$\langle \alpha_{\mathrm{c}} \rangle(t) = \int\limits_{\mathbb{R}_{\geq 0}} A p_{\alpha|n_{\mathrm{engine}}(t)}(A)\ \mathrm{d}A. \tag{8.2}$$

During transient simulations, the conditional probability density function for the heat transfer coefficient $p_{\alpha|n_{\text{engine}}(t)}$ has to be used. It strongly depends on the engine speed $n_{\text{engine}}(t)$ and the engine load. However, due to the fact that a race engine is investigated, which spends most of the time in a complete full load or coasting condition, part load conditions are neglected. For the investigated race track its percentage is less than 10 %. For the reference temperature, in this case the gas temperature T_{gas} within the combustion chamber, an analogous procedure is done. According to Eq. (3.3), the statistically modified reference temperature $T_{\text{mod}} = \langle \alpha_{\text{c}} T_{\text{gas}} \rangle / \langle \alpha_{\text{c}} \rangle$ has to be used. In the following, it is called ACT (Average Cylinder Temperature). In this chapter, the determination of $p_{\alpha|n_{\text{engine}}(t)}$ is based on the Woschni model (Woschni, 1967). Coasting conditions are treated in a similar manner.

Assuming that only two kinds of heat fluxes exist, following ordinary differential equation for the engine temperature T_{Cyl} is solved numerically:

$$C_{\text{v}} \rho \Delta x \dot{T}_{\text{Cyl}}(t) = \chi \langle \alpha_{\text{c}} \rangle (t) \left(T_{\text{mod}}(t) - T_{\text{Cyl}}(t) \right) + \alpha_{\text{w}}(t) \left(T_{\text{w}}(t) - T_{\text{Cyl}}(t) \right), \tag{8.3}$$

$$T_{\text{w}}(t) = T_{\text{w,i}} + \dot{Q}(t) / \left(\dot{m}_{\text{w}}(t) C_{\text{p,w}} \right). \tag{8.4}$$

$T_{\text{w,i}}$ is the inlet water temperature, and serves as a model input parameter. $\dot{Q}$ is the heat flux which is transferred from the solid to the water. The time-dependent variables α_{c} and α_{w} are the effective heat transfer coefficients according to Eq. (8.2) and (8.1). The parameter χ, which should be smaller than one, but larger than zero, describes the ratio between the effective surfaces of the combustion chamber and the water jacket. It serves as an additional calibration parameter, which is set to 0.3. A too large value would result in too high solid temperatures. Δx is the characteristic wall thickness, and is necessary for dimensional reasons. It can be seen as the volume of the representative engine part, which is described by a lumped capacity, divided through the effective surface area of the water wetted side. Therefore, it is not completely independent, e.g., it has to be scaled according to the existing geometry. In this thesis, a value of $\Delta x = 15$ mm is used. C_v and ρ are the heat capacity and the density of the solid, respectively. Typical values for aluminium cylinder heads are in the range of 900 J/(kgK) and 2.7 kg/m^3. Concerning the water mass flow, as already discussed in Chapter 5, oscillating frequencies in the range of $1 - 5$ Hz can increase, respectively decrease, the Nusselt number in the range of 10 %. To ensure the model validity and the technical feasibility, the water pump is modeled as a PT1-element with limited water mass flow rates. Local Reynolds numbers should be noticeably larger than 2300, the transition number between laminar and turbulent pipe flow. In addition, if the water mass flow rate is very low, the amplitudes of the water temperature will be quite large: see Eq. (8.4). Normally, the water pressure is limited and, therefore, to avoid local boiling and corresponding cavitation damage, a minimum flow rate must be guaranteed. On top of that, the design of a suitable vehicle radiator is much complicated for large oscillations, and fluctuating water inlet temperatures are undesirable for real engines. The transfer function for the water pump in Laplace space is

$$G_\mathrm{p} = \frac{1}{\tau_\mathrm{p} s + 1}, \tag{8.5}$$

with a characteristic time constant τ_p, which is set to 0.2 s, resulting in a base frequency of 5 Hz. Therefore, its dependency on the transient pressure build-up and the mechanical behavior of the pump is neglected. The allowable range for the water mass flow rate during the simulation is qualitatively shown in Fig. 8.2.

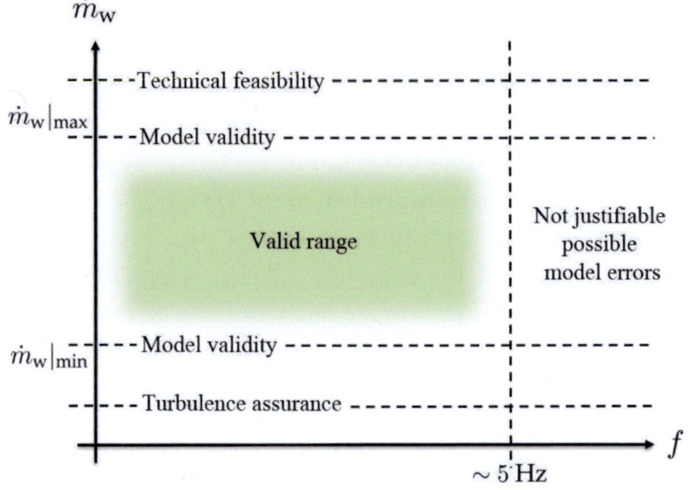

Figure 8.2: Allowable range for the water mass flow rate during the simulation in the time ($\dot{m}_{\mathrm{W}}|_{\mathrm{min}}$ and $\dot{m}_{\mathrm{W}}|_{\mathrm{max}}$) and frequency (5 Hz) domain.

8.2.3 Detailed thermal model - CFD-CHT method

The detailed finite volume model of the engine is the same as in Chapter 5, extended by a feed forward and control device. The adaption of the water reference temperature is also modeled according to Eq. (8.4). The operating point of the water pump is also limited according to Fig. 8.2. Therefore, the transfer function (8.5) is used.

8.2.4 Feed forward strategy

The block diagram for the feed forward strategy is shown in Fig. 8.3. In this case, only the upper path is active. The time-dependent disturbance $z(t)$ on the engine are transient heat fluxes caused by alternating firing and coasting conditions. The control variable is the cylinder head temperature T_{Cyl} of a representative measuring position close to the combustion chamber. The correcting variable for all scenarios is the water mass flow rate $\dot{m}_{\mathrm{w}}$. $T_{\mathrm{Cyl,t}}$ is the command variable. $\dot{m}_{\mathrm{n,w}}$ is the nominal water mass flow rate.

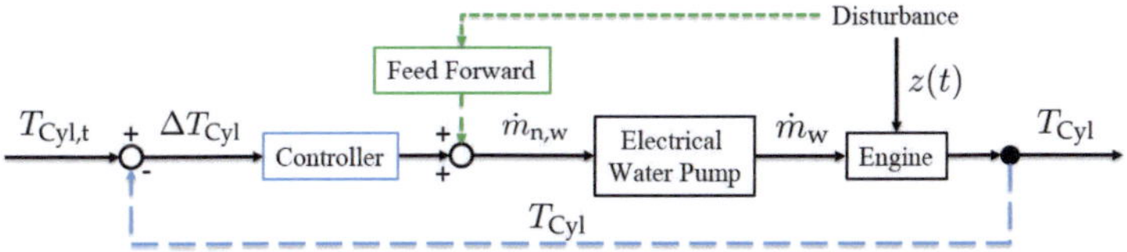

Figure 8.3: Block diagram for various feed forward and control strategies: for the pure feed forward strategy, only the upper part, visualized in green, is active. The control strategy, as can be seen with the blue, lower path, needs an additional, closed control loop.

Neglecting changes in the water temperature according to Eq. (8.4), and setting $\dot{T}_{\mathrm{Cyl}} = 0$ and $T_{\mathrm{Cyl}} = T_{\mathrm{Cyl,t}}$ in Eq. (8.3), it follows with Eq. (8.1) for the nominal water mass flow rate

$$\dot{m}_{\mathrm{n,w}}(t) = \left(\frac{\chi \langle \alpha_{\mathrm{c}} \rangle (t) \left(T_{\mathrm{mod}}(t) - T_{\mathrm{Cyl,t}} \right)}{- \left(\alpha_{\mathrm{ref}} / \dot{m}_{\mathrm{w}} |_{\mathrm{ref}}^{m} \right) \left(T_{\mathrm{w,i}} - T_{\mathrm{Cyl,t}} \right)} \right)^{1/m} . \tag{8.6}$$

8.2.5 Control strategy

The block diagram for the closed control loop is shown in Fig. 8.3. In this case, only the lower path, a closed control loop, is active. It is assumed that there are no measurement errors or delay elements in the feedback path: compared to the investigated temperature range, the error of 1 K for a typical thermocouple (type K) is very small. In addition, the time constant of such a measurement element, in relation to engine temperatures, is larger by more than one order of magnitude. The controller is executed as PID control device with following transfer function in Laplace space:

$$G_{\mathrm{c}} = k_{\mathrm{P}} + \frac{k_{\mathrm{I}}}{s} + k_{\mathrm{D}} s, \tag{8.7}$$

with k_P, k_I and k_D as the control parameters. Equation (8.7) can be rewritten in terms of two time constants $T_\mathrm{R,1}$ and $T_\mathrm{R,2}$:

$$G_\mathrm{c} = k_\mathrm{I}\frac{(1+T_\mathrm{R,1}s)(1+T_\mathrm{R,2}s)}{s}, \tag{8.8}$$

with $k_\mathrm{P} = k_\mathrm{I}(T_\mathrm{R,1}+T_\mathrm{R,2})$ and $k_\mathrm{D} = k_\mathrm{I}T_\mathrm{R,1}T_\mathrm{R,2}$. It must be noted that Eq. (8.3) is non-linear in the water mass flow rate $\dot{m}_\mathrm{w}$. Therefore, in order to model the engine as a PT1-element with a characteristic time constant τ_e and a static amplification $k_\mathrm{s,e}$, a linearization at a representative operating point is necessary. Neglecting changes in the water reference temperature according to Eq. (8.4), it follows for the engine temperature difference ΔT_Cyl

$$\underbrace{\frac{C_\mathrm{v}\rho\Delta x}{\alpha_\mathrm{w,0}}}_{=\tau_\mathrm{e}}\Delta\dot{T}_\mathrm{Cyl} + \Delta T_\mathrm{Cyl} = \frac{T_\mathrm{w,0}-T_\mathrm{Cyl,0}}{\alpha_\mathrm{w,0}}\Delta\alpha_\mathrm{w}, \tag{8.9}$$

$$= \underbrace{\frac{T_\mathrm{w,0}-T_\mathrm{Cyl,0}}{\alpha_\mathrm{w,0}}0.7\dot{m}_\mathrm{w,0}^{-0.3}\frac{\alpha_\mathrm{ref}}{\dot{m}_\mathrm{w}|_\mathrm{ref}}}_{=k_\mathrm{s,e}}\Delta\dot{m}_\mathrm{w}, \tag{8.10}$$

with $\Delta T_\mathrm{Cyl} = T_\mathrm{Cyl} - T_\mathrm{Cyl,0}$. The subscript 0 corresponds to the representative operating point. The last step is the result of Eq. (8.1). As a first approximation for the control parameters, the rule according to Kessler (Kessler, 1955) is used. Compensating both the time constants of the water pump τ_p and the engine τ_e, the transfer function for the open control loop G_o and the closed loop G_cl in Laplace space is

$$G_\mathrm{o} = \frac{k_\mathrm{I}k_\mathrm{s,e}}{s}, \tag{8.11}$$

$$G_\mathrm{cl} = \frac{G_\mathrm{o}}{1+G_\mathrm{o}} = \frac{1}{1+\frac{1}{k_\mathrm{I}k_\mathrm{s,e}}s}. \tag{8.12}$$

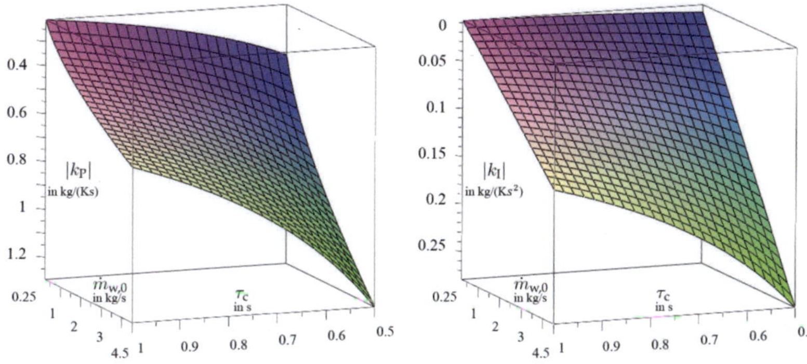

(a) (b)

Figure 8.4: First approximations for the control parameters k_P and k_I of the PID control device as functions of the water mass flow rate in the representative operating point $\dot{m}_{\mathrm{w},0}$ and the resulting system time constant of the closed control loop $\tau_\mathrm{c} = 1/(k_\mathrm{I} k_{\mathrm{s,e}})$.

As an example, Fig. 8.4 shows the k_P and k_I parameters of the PID control device for a representative engine operating point. The given values should be regarded as first, possible approximations because of the strong non-linearity and the ever changing engine states. According to Eq. (8.9) and (8.10), representative temperatures $T_{\mathrm{w},0} = 373$ K and $T_{\mathrm{Cyl},0} = 419$ K are used for the time constant τ_e and the amplification $k_{\mathrm{s,e}}$. The control parameters are given as functions of the water mass flow rate in the representative operating point $\dot{m}_{\mathrm{w},0}$ and the resulting system time constant of the closed control loop $\tau_\mathrm{c} = 1/(k_\mathrm{I} k_{\mathrm{s,e}})$.

8.3 Results and discussion

8.3.1 Feasible temperature range and consistency

Firstly, the feasible temperature range and the temporal behavior is investigated by means of the lumped capacity model. Therefore, the minimum water mass flow rate is set to 0.25 kg/s, whereas the maximum rate is 4.5 kg/s. As can be seen in Fig. 8.5 a), a temperature range about 80 K can be realized by using a water pump with a variable water mass flow rate. The limitations result from the allowable parameter space in Fig. 8.2. However, as can be seen in Fig. 8.5 b), the standard deviation of the temperature rises strongly with higher target temperatures. Allowing a maximum value of 6 K, which can also be reached by a conventional mechanical water pump, a range of 60 K can be realized. The reasons are the same as for fig 8.5 a): targeting a component temperature of 380 K, the resulting water mass flow rate is constant at 4.5 kg/s. One gets exactly the opposite result for a target temperature of 470 K, e.g., a nearly constant mass flow rate of 0.25 kg/s. However, the temporary, local temperature minima are deeper because of larger thermal differences between firing and coasting engine conditions.

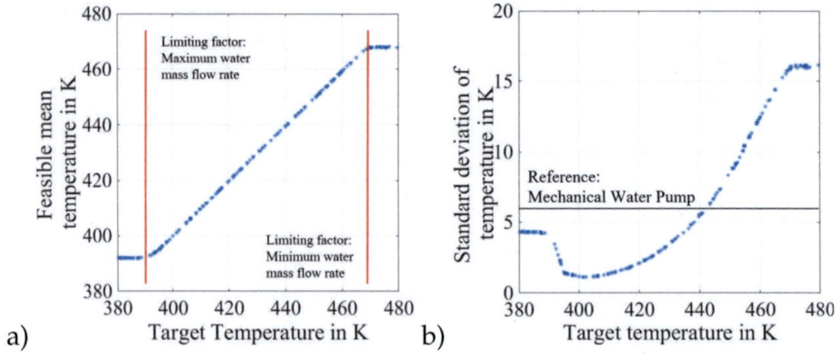

Figure 8.5: Lumped capacity model: feasible temperature range by using a PID controller with $k_P = -1.4$ kg/(Ks), $k_I = -0.05$ kg/(Ks^2) and $k_D = -1.0$ kg/K. a) Mean temperature. b) Standard deviation of the temperature.

8.3.2 Monte Carlo simulation

Fig. 8.6 to 8.8 show some results of a Monte Carlo simulation for a constant target temperature of 407 K. The results in Fig. 8.4 serve as starting values. Regarding regular temperature curves under a transient drive, high k_P and small k_I absolute values are expedient, see Fig. 8.6. However, k_P values higher than 0.6 do not provide a significant advantage if the k_I value is small enough. Because open loops without integrating parts do not lead to stationary accurate closed loops, k_I levels close to zero are not recommended.

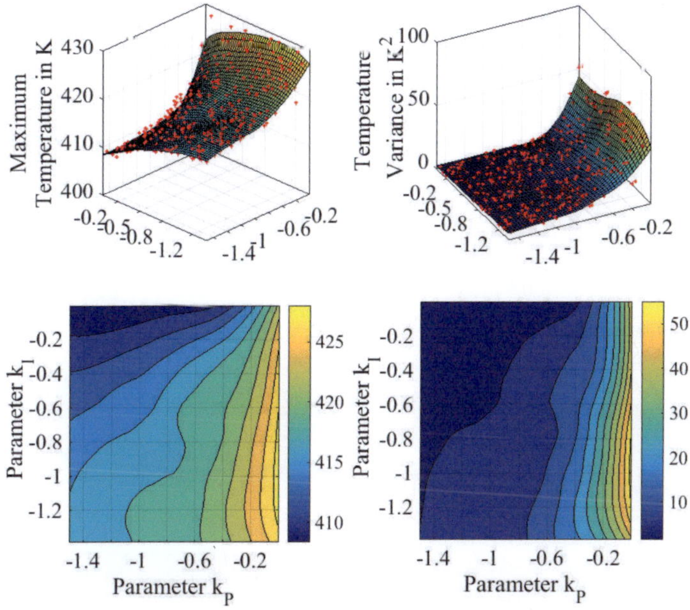

Figure 8.6: Lumped capacity model: Monte Carlo simulation for a target temperature of 407 K in the k_P - k_I space. The resulting cylinder head temperature is shown.

The limit theorem for Laplace-Transformations

$$\lim_{t \to \infty} y_l(t) = \lim_{s \to 0} sY_l(s), \tag{8.13}$$

for a function y_l and its Laplace transform Y_l, requires following condition for stationary accuracy:

$$G_{cl}(0) = \frac{G_o(0)}{1 + G_o(0)} = 1. \tag{8.14}$$

This can be only fulfilled with $\lim_{s \to 0} G_o(s) \to \infty$. Therefore, k_I values different to zero are necessary. Values about -0.05 and -0.1 are recommended. In the k_P - k_D space, a similar behavior can be observed: high k_P and k_D absolute values are expedient. With values higher than 1.2 and 0.3, no significant advantages can be observed. In general, the influence of the k_D parameter is low. Regarding hydraulic power losses, a similar result can be observed. Fig. 8.8 shows the example of the k_P - k_I space. Concerning a holistic consideration, e.g., taking into account the maximum hydraulic power, its variance and mean value, a similar range in the k_P - k_I - k_D space is recommended.

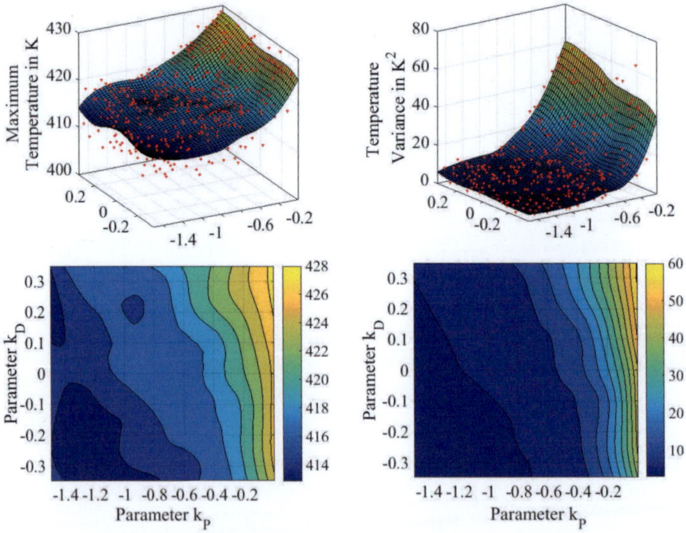

Figure 8.7: Lumped capacity model: Monte Carlo simulation for a target temperature of 407 K in the k_P - k_D space. The resulting cylinder head temperature is shown.

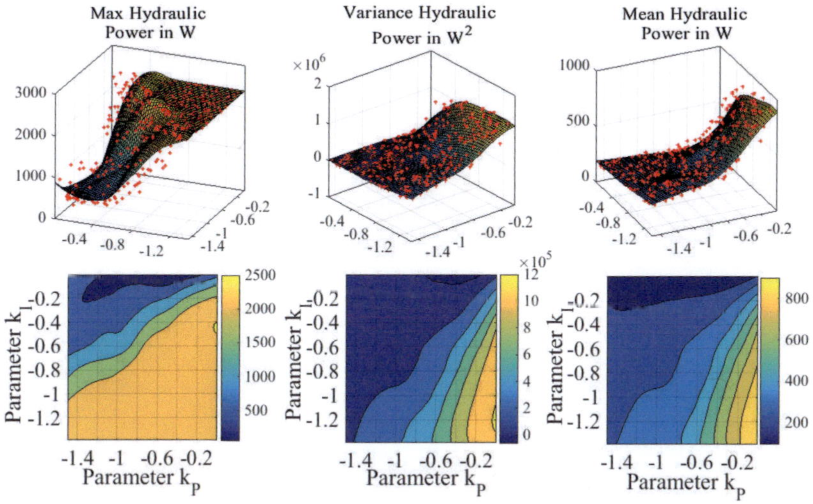

Figure 8.8: Lumped capacity model: Monte Carlo simulation for a target temperature of 407 K in the k_P - k_I space. The resulting hydraulic power is shown.

8.3.3 Dynamic effects

On the comparability of the following strategies, the boundary condition for all cases is the same: a fixed maximum engine temperature under full load state at the end of straight track. As can be seen in Fig. 8.9, the pure feed forward strategy decreases significantly the temperature standard deviation. Especially in coasting situations or in the beginning of acceleration phases, the temperature curve is smoother. While accelerating out of corners, a further advantage can be gained if a controller is used simultaneously. Using a higher integrating and no differentiating part for the PID controller, a certain temperature overshoot can be realized, which results, with regard to a mechanical water pump, in a slightly higher heat saving. According to the lumped capacity model, a heat saving potential about one kilowatt is possible. This corresponds to a share of approximately 2 % of the total heat loss.

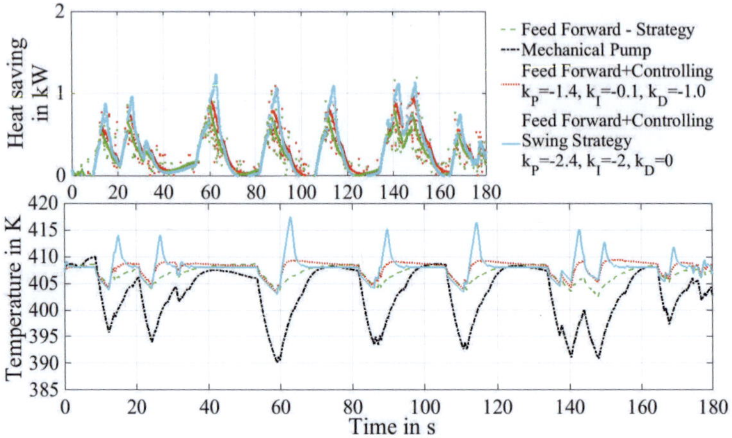

Figure 8.9: Lumped capacity model: comparison of four different feed forward and control strategies concerning heat saving potentials and temperature consistency under a transient drive.

8.3.4 Temperature shifting and resulting heat saving

Next, different target temperatures are used and compared to a mechanical water pump. As can be seen in Fig. 8.10, according to the lumped capacity model, a heat saving potential of five kilowatts can be achieved by shifting the temperature up to 470 K.

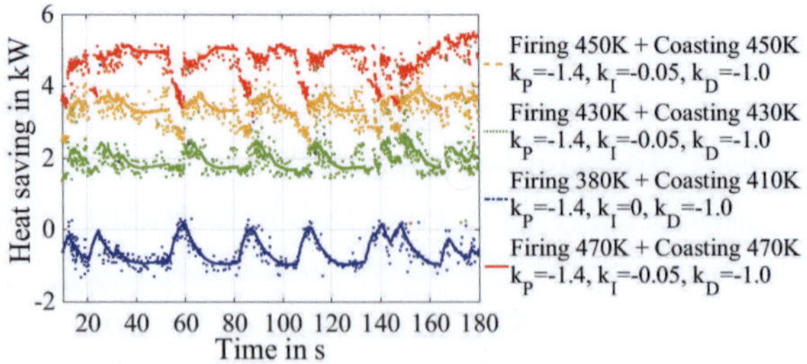

Figure 8.10: Lumped capacity model: comparison of various control strategies and target temperatures concerning heat saving potentials and temperature consistency under a transient drive.

This corresponds to a share of approximately 10 % of the total heat loss. Interestingly, a very smooth temperature curve can be realized if one uses anticyclical target temperatures, e.g., a higher target temperature under coasting conditions than under fired situations.

8.3.5 Three-dimensional results

In the following, according to Fig. 8.1, transient finite volume simulations are presented and compared with the lumped capacity model.

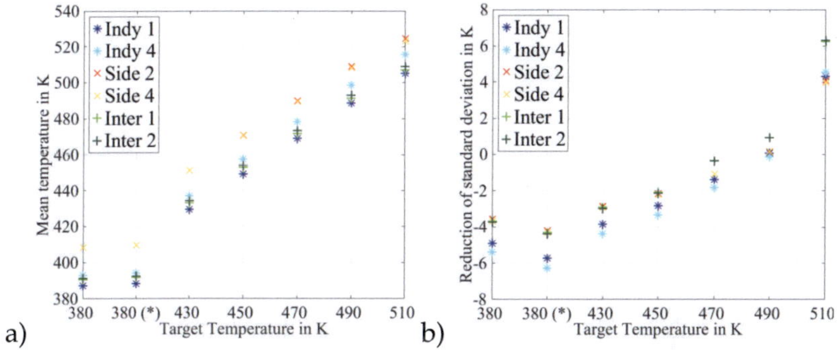

Figure 8.11: Detailed 3D simulation: feasible temperature range by using a PID controller with $k_P = -1.4$ kg/(Ks), $k_I = -0.05$ kg/(Ks^2) and $k_D = -1.0$ kg/K. a) Mean temperature. b) Mean differences in the temperature standard deviation with regard to a mechanical pump. The ratio of the (reference) mechanical water pump is set in order to reach a maximum temperature of 407 K.

Compared to Fig. 8.5, mean temperature results, together with their standard deviations, for the detailed FVM simulation are shown in Fig. 8.11. In case of the detailed simulation, one can notice a larger temperature range for the feasible mean temperature. Temperature values up to 510 K are possible. One possible reason can be the additional heat flux due to heat conduction. It seems that, with higher mean temperatures, its relative weighting falls in relation to the heat flux due to water cooling. This can be explained with the higher overall temperature level. As a consequence, the water cooling becomes more important, e.g., the temperature difference between solid and water has to be larger. With regard to Fig. 8.5 b) with Fig. 8.11 b), one can notice a smaller sensitivity with regard to the temperature variance. One argument can also be the additional heat flux due to heat conduction, and, therefore, the lower sensitivity towards water mass flow rates. In addition, the temperature gradients in the detailed simulation cause a kind of filtering function. In case of the lumped model, there exist no spatial gradients, and, therefore, every change in the boundary conditions directly affects the solid temperature. The measurement positions are the same as in Fig. 5.8. Again, the numbers indicate different cylinders and blocks. In this case, the command variable is related to the position "Indy 1". The second setting which is called 380 K (*) corresponds to the anticyclical target temperature strategy from Fig. 8.10. A key element appears to be that the other five measurement points near the combustion chamber behave in a manner similar to that of the command variable "Indy 1".

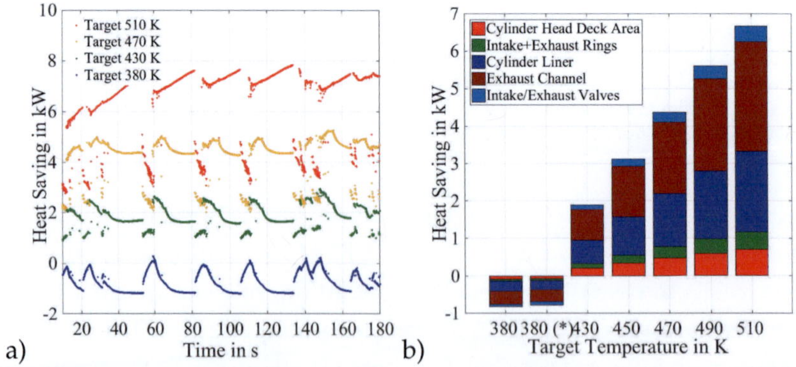

Figure 8.12: Detailed 3D simulation: a) Heat saving potential with regard to a mechanical pump. The PID controller parameters are the same as in Fig. 8.11. a) Transient heat saving potential. b) Mean heat saving potential with corresponding spatial distribution. The ratio of the (reference) mechanical water pump is set in order to reach a maximum temperature of 407 K.

Compared with Fig. 8.10, transient heat saving potentials for the detailed simulations are shown in Fig. 8.12. Under dynamic operating conditions, heat savings up to 8 kW are possible for a target temperature of 510 K. This corresponds to a share of approximately 16 % of the total heat loss. Because of the large standard deviation at such high temperature levels, this potential drops below 4 kW under coasting conditions. As can be seen in Fig. 8.12 b), most of the heat is saved in the exhaust channels and the cylinder liners. Surprisingly, the cylinder head deck area has a small proportion. The higher heat saving potential, in case of the detailed simulation, is a result of the higher, feasible mean temperatures.
Compared to Fig. 8.9, the detailed FVM simulations show a higher transient heat saving potential in acceleration phases. As can be seen in Fig. 8.13, the temperature swing strategy delivers values up to 2 kW. Regarding the integrating part of the controller, it is important to note, however, that the oscillation behavior is more pronounced.

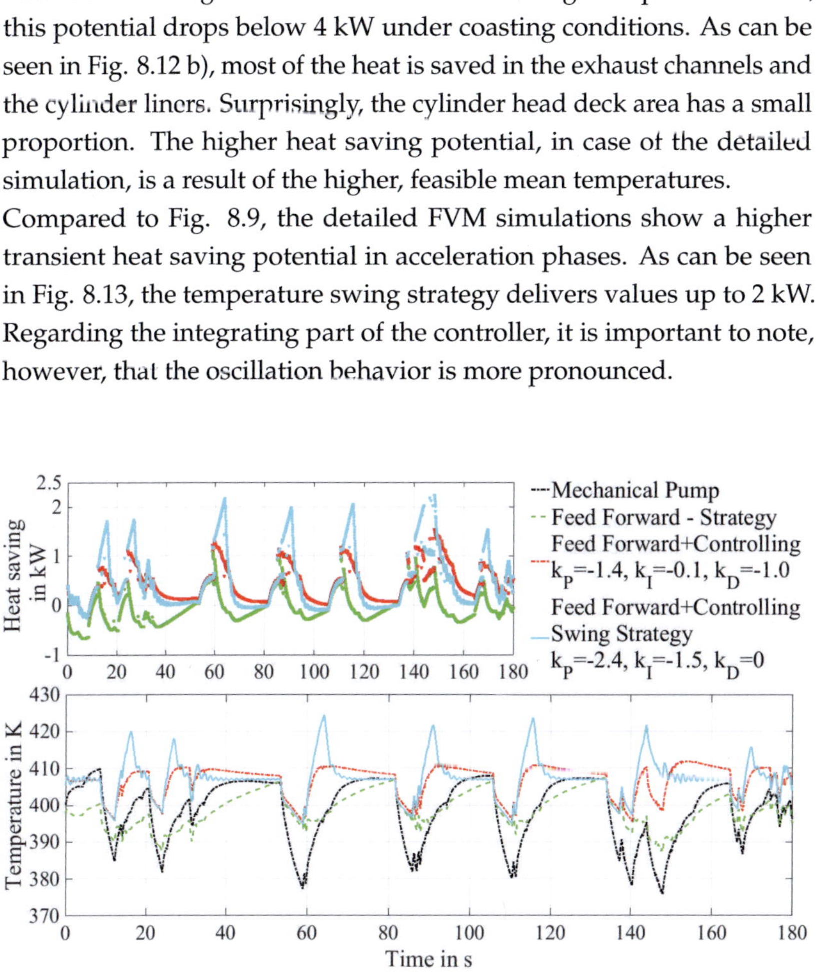

Figure 8.13: Detailed 3D simulation: comparison of four different feed forward and control strategies concerning heat saving potentials under a transient drive.

Interestingly, with respect to the lumped capacity model, the difference between the pure feed forward strategy and the combination of pre-control and feedback control is higher. In general, the lumped capacity model underestimates the temperature swings: the lower temperatures during coasting are overestimated in the lumped model. One possible explanation can be the missing heat flux due to heat conduction, which is only contained in the three-dimensional simulation. As a result of this lower temperature difference between the mechanical pump and the control strategy, the heat saving potential is higher for the detailed simulation. In addition, the real, inhomogeneous distribution of the heat flux vector around the combustion chamber cannot be realized with a lumped model approach.

8.4 Conclusions

Concerning the initial research questions, the results are as follows. Within the limits of 0.25 and 4.5 kg/s for the water mass flow rate, a feasible range for the mean temperatures of engine components near the combustion chamber is 130 K. Compared with a conventional mechanical water pump, the standard deviation can be reduced up to 6 K. For the highest mean temperature, the standard deviation is 6 K higher. A very important finding is the fact that, by using one command variable, the other measurement points near the combustion chamber behave in a similar manner. Concerning heat fluxes in full load states, it could be shown that a heat saving potential about 8 kW is possible. This corresponds to a share of approximately 16 % of the total heat loss. In re-acceleration phases, values about 2 kW are realistic. Most of the heat losses can be reduced in the exhaust channels and the cylinder liners: approximately 75 % of the complete heat flow is reduced in this area. Of the remaining 25 %, half of the heat is reduced at the cylinder head deck area. Valves and their rings contribute equally to the rest.

Concerning the two different modelling ideas, the qualitative progressions of relevant physical quantities are very similar. Therefore, the lumped capacity model can be used for optimization and sensitivity calculations: the Monte Carlo simulation has shown valuable correlations for the control device parameters, which could be effectively used in subsequent three-dimensional simulations. Compared to these detailed simulations, the feasible range for the mean temperatures is underestimated by 50 K. Analogously, the profit concerning standard deviations is slightly underestimated by 1 K. However, for high temperature levels, it is overestimated by 6 K. With regard to heat saving potentials, the simplified model gives too low values for both the dynamic and the quasi-stationary case: In the first one, the difference is 3 kW, in the second one, it is 1 kW.

Chapter 9

Conclusions and outlook

Concerning the overall research question of this thesis what might a suitable calculation method for the temperature and fatigue behavior look like, the results are as follows. With the help of in-cylinder pressure measurements, using focused model reduction, probability density functions, linearization and extrapolation, respectively, the cycle-to-cycle variability can be considered without detailed, crank angle resolved simulations. Thereby, the presented stationary, as well as transient, temperature calculation method accounts for different thermal behaviors of various engine components, cylinder-individual temperature effects, as well as the transient heat transfer within the water jacket. Concerning the influence of various engine settings, however, one has to pay attention which model in detail is used. Furthermore, the selection should be different for stationary and transient applications. In case of transient applications, with specific temperature increases of approximately 1 K/kW for the cylinder head, and 2 K/kW for the cylinder liner, the influence of the air-fuel ratio is quite strong. Concerning the influence of various engine settings on the lifetime of engine pistons under dynamic operating conditions, the specific results are qualitatively similar to the temperature differences. With a safety factor change of approximately 1.5 - 2.5 %/kW, the air-fuel ratio has the strongest influence.
Due to some deficits for ignition time variations, presented in Chapter 3 for stationary investigations and Chapter 4 for the part load model, further investigations concerning more advanced models like Bargende (1991) should be made. In general, in case of the part load model, it

could be worthwhile modeling further effects like the turbulent intensity enhanced by spray (Grill, 2006). In addition, research should be more focused on modeling of the correlation between the heat transfer coefficient and the gas temperature. Typical applications are series-production engines with a wide range of part load states: the proposed method can benefit advanced ECU (Electronic Control Unit) functionalities which aim to predict thermal loads under a transient drive. In this sense, concerning the application within the transient calculation method from Chapter 5, further investigations with a wider range of the engine load should be conducted. In addition, the determination of the Reynolds exponent in Section 5.5.4 can be improved by using advanced heat transfer models for thermal entrance regions. Analogously, the transient heat transfer modeling during gear changes could take into account oscillating frequencies and amplitudes. Furthermore, concerning the spatial distribution of the heat fluxes within the combustion chamber, preconnected, crank angle resolved CFD simulations will ensure a wider applicability than the assumed progressions from Chapter 2. However, due to the satisfied simulation of transient component temperatures, with regard to more realistic boundary conditions, their influence within stationary, state-of-the-art 1D modeling tools should be interesting with regard to knocking modeling and volumetric efficiency. Similarly, they can be used for the optimization of a water jacket design and the evaluation of local heat fluxes.
Motivated by several consistencies between simulation and experiment for the thermodynamic variation under a transient drive, presented in Chapter 6, enhancements should be derived in case of fuel flow- and exhaust gas variations. For the first one, alternative heat transfer models are necessary. In case of the second one, the effect of transient heat fluxes on the exhaust gas, as well as chemical kinetics, could lead to some improvements.
With regard to the piston failure analysis in Chapter 7, further investigations concerning model validation should be made with the help

of statistical methods. In addition to the scattering nature of the combustion process, the scattering material properties make a validation difficult. One of the most important outstanding research questions is the additional damage effect of knocking cycles and low cycle fatigue loading. The first one is characterized by increased pressure values and additional heat fluxes, as well as by pressure waves and their local causation of vibrations and destructions of the oil film. The second one can be superimposed by temporary cooling and heating cycles, e.g., thermomechanical fatigue. However, this effect depends strongly on the engine application and the investigated points. As an example, pronounced thermomechanical fatigue should be observed around the spark plug within the cylinder head. For higher temperature applications, the additional damage of creep is important. To summarize, the question of damage accumulation should be addressed. Therefore, micromechanics could be a key to success. As reported in Myers (1990), the dominant damage mechanism can change, and, depending on the material, crack growth could be more relevant. Furthermore, possible effects of temperature swings on the piston surface should be considered. Using damage maps, the proposed method could benefit advanced ECU functionalities which aim to predict material fatigue, as well as damage monitoring, under a transient drive. This approach could also be of interest to other engine components. In this context, depending on the material, the influence of damage modeling, like the stress localization and microplasticity law, or the crack initiation condition with the corresponding crack growth model, should be investigated. From the point of micromechanics, local influences of the grain size and orientation, different thermal treatments and defects, referred to various manufacturing processes, will significantly determine the fatigue behavior.

In relation to Chapter 8, the direct influence on the combustion, either positively affected by reduced heat losses, and stabilized temperature conditions, or negatively affected by higher inlet air temperatures, resulting in different knocking tendencies and volumetric efficiencies, has to be investigated experimentally. In addition, the model validity and its scope of application, shown in Fig. 8.2, should be extended by the laminar transition regime, as well as by higher frequencies, and also validated through different perspectives of control strategies.

Frequently Used Acronyms, Symbols, and Operators

Acronyms

ACT	Average Cylinder Temperature
BC	Boundary Condition
CA	Crank Angle
CFD	Computational Fluid Dynamics
CHT	Conjugate Heat Transfer
DOE	Design of Experiments
ECU	Electronic Control Unit
FEM	Finite Element Method
FVM	Finite Volume Method
HTC	Heat Transfer Coefficient
IVC	Inlet Valve Closing
PDF	Probability Density Function
PID	Proportional Integral Derivative
RANS	Reynolds Average Navier Stokes
SI	Spark-Ignition
TDC	Top Dead Center

Latin letters

a	Temperature conductivity $a = \lambda / (\rho c_p)$
a_t	Turbulent temperature conductivity
a_1	Model parameter turbulence modelling
A	Realisation of random variable α
A_eff	Effective area
A_i	Fundamental quantity
$\boldsymbol{b}$	Acceleration vector
$b_\mathrm{T} - e_\mathrm{T}$	Model constants (metal fatigue) for temperature dependency
B	Engine bore
c_p	Specific heat at constant pressure
c_v	Specific heat at constant volume
$C_\mathrm{p,w}$	Specific heat at constant pressure for water
$C_1 - C_3$	Model parameters
d_0	Jet nozzle diameter
d_s	Diameter of a sphere with the same volume as the instantaneous cylinder volume
D	Damage state
$\tilde{D}$	Representative damage state
D_hyd	Hydraulic diameter
D_v	Valve diameter
D_ω	Production term
e	Specific inner energy
$edges$	Matrix which represents $\underline{M}$
E	Error between simulation and experiment
$\tilde{E}$	Empirical expectation value of error E
E_κ	Von-Karman constant
f	Frequency
$\tilde{f}$	Arbitrary random variable

f_{slave}	Approximated inner integral according to Eq. ((5.5))
f_{master}	Argument of outer integral according to Eq. ((5.5))
$\boldsymbol{F}$	Lateral force on piston skirt
F_1	Smoothing function
F_2	Smoothing function
g	Gravity
$\tilde{g}$	Smothing function
G_{c}	Transfer function for the controller
G_{cl}	Transfer function for the closed control loop
G_k	Production term
G_{o}	Transfer function for the open control loop
G_{p}	Transfer function for the water pump
Gr	Grashof number
Gz	Graetz number
G_ω	Production term
H_i	Thickness of layer i
$\boldsymbol{I}$	Second order identity tensor
k	Turbulent kinetic energy
$\tilde{k}$	Boltzmann constant
k_{D}	D-component of the controller
k_{I}	I-component of the controller
k_{P}	P-component of the controller
$k_{\mathrm{s,e}}$	Static amplification engine
k_{W}	Inverse slope of double logarithmic Woehler line
l	Characteristic length
l_{v}	Valve lift
m	Re exponent
m_{air}	Mass flow of air
m_{fuel}	Mass flow of fuel
$\dot{m}_{\mathrm{n,w}}$	Nominal value of the water mass flow rate

m_p	Mass of particle (atom, molecule)
$\dot{m}_{\mathrm{w}}$	Water mass flow rate
$\dot{m}_{\mathrm{w}}\|_{\mathrm{ref}}$	Reference water mass flow rate
$\dot{m}_{\mathrm{w}}\|_{\mathrm{max}}$	Maximum water mass flow rate
$\dot{m}_{\mathrm{w}}\|_{\mathrm{min}}$	Minimum water mass flow rate
m_{water}	Mass flow within the water channel
$\underline{M}$	Five-dimensional engine state matrix of outer boundary conditions
$\underline{M}_{\mathrm{stat}}$	Stationary engine state matrix of outer boundary conditions for reference purposes
$\underline{M}_1$	Five-dimensional engine state matrix: Given reference state (full load)
$\underline{M}_2$	Five-dimensional engine state matrix: Given state (part load)
n	Pr exponent
$\underline{n}$	5D random variable describing the engine state
$\boldsymbol{n}$	Boundary normal vector
n_{engine}	Engine speed
n_{Sim}	Matrix row of pointer matrix P for a given simulation time t_{Sim}
N	Amount of substance
$\underline{N}$	Realisation of random variable $\underline{n}$
N_{c}	Cycle number
$N_{\mathrm{c},k}$	Cycle number for pressure cycle k
$N_{\mathrm{c,t}}$	Target cycle number
Nu	Nusselt number $\alpha l/\lambda$
Nu_{v}	Nusselt number $\alpha l_{\mathrm{v}}/\lambda$
p	Static pressure
p_{cmb}	Cylinder pressure due to combustion
p_{hp}	Pressure of high-pressure cycle
p_{ini}	Initial pressure
p_{int}	Boost pressure

p_{lp}	Pressure of low pressure cycle
p_{m}	Motored cylinder pressure
p_{max}	Maximum pressure within one engine cycle
p_n	Probability density function on n
$p_{n\vert\underline{M}}$	Conditional probability density function on n with regard to $\underline{M}$
$p_{n\underline{M}}$	Joint probability density function on n and $\underline{M}$
$\hat{p}_n$	Normed histogram; Approximated probability density function on n
P	Pointer matrix
Pr	Prandtl number ν/a
Pr_{t}	Turbulent Prandtl number $\nu_{\mathrm{t}}/a_{\mathrm{t}}$
q_{wall}	Wall heat flux
$\boldsymbol{q}$	Heat flux vector
$\boldsymbol{q}_{\mathrm{cond}}$	Heat flux vector due to heat conduction in the solid
$\dot{Q}$	Heat flux
Q_j	Dimensioned quantity
r	Rank of dimension matrix
r_i	Radius of layer i
R	Universal gas constant
$\boldsymbol{R}$	Reynolds stress tensor
Re	Reynolds number lv/ν
Re_{j}	Exhaust jet Reynolds number lv_{j}/ν
Re_0	Oil jet Reynolds number d_0v_0/ν
Re_{v}	Reynolds number $l_{\mathrm{v}}v_{\mathrm{j}}/\nu$
R_i	Thermal resistance of layer i
R_{s}	Stress ratio
s	Laplace coordinate
S	Modulus of the mean strain rate tensor $S = \sqrt{2\boldsymbol{S}\cdot\cdot\,\boldsymbol{S}}$ with $\boldsymbol{S} = \mathrm{sym}(\mathrm{grad}\,(\langle\boldsymbol{u}\rangle))$
S_f	Safety factor

t	Physical time
t_{amb}	Ambient temperature of air
t_{int}	Inlet temperature of air
t_{Sim}	Simulation time
T	Temperature
$\tilde{T}$	Turbulent time scale
T^{+}	Dimensionless temperature
T_{Cyl}	Cylinder head temperature
$T_{\mathrm{Cyl,t}}$	Command variable of the cylinder head temperature
$\overline{T}_{g}$	Cylinder-average gas temperature
T_{gas}	Reference (gas) temperature
T_{i}	Indicated torque by combustion
T_{mod}	Statistically modified temperature
$T_{\mathrm{o,i}}$	Inlet oil temperature
T_{ref}	Reference temperature of fluid
$T_{\mathrm{R,1}}$	Time constant of control device
$T_{\mathrm{R,2}}$	Time constant of control device
T_{s}	Solid temperature
T_{ub}	Unburnt gas temperature
T_{w}	Water reference temperature
T_{wall}	Wall temperature
$T_{\mathrm{w,i}}$	Inlet water temperature
$T_{\mathrm{w,o}}$	Outlet water temperature
u	Realisation of random variable U
$\tilde{u}$	Realisation of random variable $\tilde{U}$
u^{+}	Dimensionless velocity
u^{*}	Friction velocity
$\boldsymbol{u}$	Velocity vector
U	Arbitrary random variable
$\tilde{U}$	Arbitrary random variable
v	Characteristic velocity

v_{c}	Scaled combustion convection
v_{c1}	Part one of characteristic velocity according to Woschni
v_{c2}	Part two of characteristic velocity according to Woschni
v_{j}	Exhaust jet velocity through valve opening
v_0	Oil jet velocity at nozzle exit
v_{p}	Current piston speed
v_{piston}	Mean piston speed
v_{water}	Characteristic velocity within the water channel
V	Volume
V_{d}	Displaced volume
V_{max}	Maximum volume
x	Ratio between burnt mass and complete in-cylinder mass
Δx	Characteristic wall thickness
$\boldsymbol{x}$	Position vector
y	Dimensioned wall distance
y^+	Dimensionless wall distance
y_b	Ratio between burnt volume and complete in-cylinder volume
y_{l}	Arbitrary function
Y_{l}	Laplace transform of y_{l}
z	Disturbances of all kinds on the engine
$\tilde{z}$	Nozzle to piston distance
z_d	Number of dimensioned quantities

Greek letters

α	Heat transfer coefficient
$\alpha_{\mathrm{a}}, \alpha_{\mathrm{b}}, \alpha_{\mathrm{c}}$	Model constants for metal fatigue

α_{c}	Effective heat transfer coefficient for the combustion chamber
α_{cr}	Crank angle
α_{ign}	Ignition crank angle
$\tilde{\alpha}_{i,j}$	Exponent describing the correlation between dimensioned quantity Q_j and fundamental quantity A_i
α_{l}	Model parameter for metal fatigue
α_{ref}	Reference heat transfer coefficient
α_{stat}	Heat transfer coefficient for stationary engine state $\underline{M}_{\mathrm{stat}}$
α_{w}	Effective heat transfer coefficient for the water channel
α_1	Realisation of α in the state $\underline{M}_1$
α_2	Realisation of α in the state $\underline{M}_2$
β	Model parameter turbulence modelling
β^*	Model parameter turbulence modelling
$\tilde{\beta}$	Coefficient of thermal expansion at constant pressure
$\beta_{\mathrm{a}} - \beta_{\mathrm{d}}$	Model constants for metal fatigue
β_{l}	Model parameter for metal fatigue
$\tilde{\beta}_{\mathrm{l}}$	Model parameter for metal fatigue without temperature dependency
β_T	Transformation coefficient
$\tilde{\beta}_{\mathrm{T}}$	Model parameter for metal fatigue with temperature dependency
γ	Model parameter turbulence modelling
γ_{c}	Ratio of heat capacities
$\gamma_{\mathrm{p,ac}}$	Accumulated plastic shear strain
ε	Turbulent dissipation
ε_{c}	Model constant for turbulent dissipation
λ	Thermal conductivity

λ_{cmb}	Ratio between actual air mass and stoichiometric air mass
θ	Spherical coordinate
κ	Isentropic exponent
κ_k	Logarithmic wall constant
μ	Dynamic viscosity
μ_{t}	Turbulent dynamic viscosity
ν	Kinematic viscosity μ/ρ
ν_{t}	Turbulent kinematic viscosity μ_{t}/ρ
ξ	Expectation value pressure
ξ_s	Slip direction angle
ρ	Mass density
σ	Standard deviation of pressure
$\boldsymbol{\sigma}$	Cauchy stress tensor
$\tilde{\sigma}_E$	Empirical standard deviation of error E
$\boldsymbol{\sigma}_{hp}$	High-pressure stress tensor
σ_k	Turbulent Prandtl number for k
$\boldsymbol{\sigma}_l$	Lower stress tensor
$\boldsymbol{\sigma}_{lp}$	Low pressure stress tensor
$\boldsymbol{\sigma}_{\mathrm{p}}$	Residual stress tensor due to plasticity
σ_{Pa}	Particle diameter (atom, molecule)
σ_{ref}	Reference stress
$\boldsymbol{\sigma}_{\mathrm{u}}$	Upper stress tensor
σ_ω	Turbulent Prandtl number for ω
$\langle\boldsymbol{\tau}_{\mathrm{a}}\rangle$	Macroscopic, resolved shear stress amplitude
τ_{c}	Time constant of closed loop
τ_{e}	Time constant of engine
τ_{p}	Time constant of water pump
τ_{wall}	Wall shear stress
ϕ	Spherical coordinate
χ	Ratio between effective surfaces of the combustion chamber and the water jacket

ω	Turbulent frequency $\omega = \varepsilon/k$
$\Omega^{(2,2)*}$	Reduced collision integral

Operators

$\mathrm{grad}\,(\cdot)$	Gradient: $\dfrac{\partial\,(\cdot)}{\partial u^k} \otimes \boldsymbol{g}^k$ with contravariant component u^k and corresponding reciprocal basis $\boldsymbol{g}^k$
$\mathrm{div}\,(\cdot)$	Divergence: $\mathrm{grad}\,(\cdot) \cdot\cdot \boldsymbol{I}$
$\boldsymbol{a} \cdot \boldsymbol{b}$	Scalar product $a_i b^i$ with covariant component a_i and contravariant component b^i
$\boldsymbol{A} \cdot\cdot \boldsymbol{B}$	Scalar product $A_{ij} B^{ij}$ with covariant component A_{ij} and contravariant component B^{ij}
$\boldsymbol{a} \otimes \boldsymbol{b}$	Dyadic product between two vectors $\boldsymbol{a}$ and $\boldsymbol{b}$ defined by its action on an arbitrary vector $\boldsymbol{c}$: $\boldsymbol{a} \otimes \boldsymbol{b}\,(\boldsymbol{c}) = (\boldsymbol{b} \cdot \boldsymbol{c})\,\boldsymbol{a}$
$\mathrm{sym}(\boldsymbol{A})$	Symmetric part of a tensor $\boldsymbol{A}$ according to $\mathrm{sym}(\boldsymbol{A}) = 1/2\left(\boldsymbol{A} + \boldsymbol{A}^{\mathsf{T}}\right)$
$\boldsymbol{A}^{\mathsf{T}}$	Transposed tensor of tensor $\boldsymbol{A}$ defined by: $\boldsymbol{a} \cdot (\boldsymbol{A}^{\mathsf{T}}\boldsymbol{b}) = \boldsymbol{b} \cdot (\boldsymbol{A}\boldsymbol{a})$ with vectors $\boldsymbol{a}$ and $\boldsymbol{b}$
$\mathrm{tr}(\boldsymbol{A})$	Trace of a tensor $\boldsymbol{A}$ with $\mathrm{tr}(A) = \boldsymbol{A} \cdot\cdot \boldsymbol{I}$
$\lVert\boldsymbol{a}\rVert$	Euclidean norm of a vector $\boldsymbol{a}$
$\dfrac{\mathrm{d}\,(\cdot)}{\mathrm{d}t}$	Time derivative
$\dfrac{\partial\,(\cdot)}{\partial t}$	Local time derivative
$\langle\cdot\rangle$	Expectation value with respect to time
$(\cdot)'$	Fluctuation value with respect to time

Appendix A

Pseudo code for modeling heat transfer coefficients

```
% Pseudo code for modeling the heat transfer coefficient (HTC)
% Given are {M_1, M_2, xi_rel(1xN), hist_rel_norm_Cycle(4xN)}
%           xi_rel: Vector of different HTC's
%           hist_rel_norm_Cycle: Normed histograms
%           (Approximation of the probability density function)
for i=1:1:4 % For loop over the four engine strokes
    % Calculation of the transformed realisations of random variables
    f_tilde_Cycle(i,:)=f_tilde_create(M_2,M_1,xi_rel,i);
    f_integration_Cycle(i,:)=hist_rel_norm_Cycle(i,:) ...
        .*f_tilde_Cycle(i,:);
    % Calculation of HTC expectation values (Equation (3))
    f_tilde_Expectation=f_tilde_Expectation+0.25*(...
        trapz(xi_rel,f_integration_Cycle(i,:)));
end
```

```

function [f_tilde]=f_tilde_create(M_1,M_2,xi_rel ,
    Cycle)
    m_used=0.8; % Reynolds Exponent
    % Reference and new State
    m_air_ref=M_1(1,2); m_air=M_2(1,2);
    t_int_ref=M_1(1,3); t_int=M_2(1,3);
    trq_ref=M_1(1,4); trq_cmb=M_2(1,4);
    m_fuel_ref=M_1(1,5); m_fuel=M_2(1,5);
    % Calculation of the amount of substances
    [N_before,N_after]=Composition(m_air,m_fuel);
    [N_before_ref,N_after_ref]=Composition(m_air_ref
        ,m_fuel_ref);

    f_tilde=xi_rel;
    if (Cycle==1) || (Cycle==2)
      % Straight forward
    elseif Cycle==3 % Third engine stroke
      % Motored pressure
      p_ratio_mot=N_before/N_before_ref;
      p_ratio_mot=p_ratio_mot*t_int/t_int_ref;
      % Pressure due to combustion
      p_ratio_cmb=trq_cmb/trq_ref;
      % Ratios of complete pressure
      % "p_mot" and "p_cmb"
      % should be an approximation to Fig. 4 a):
      % (1xN) matrix
      p_ratio=p_mot*p_ratio_mot...
          + p_cmb*p_ratio_cmb;
      % Amount of substances
      Mol_before=N_before/N_before_ref;
      Mol_after=N_after/N_after_ref;
```

```
            % Ratio temperature in Cycle3
            T_ratio=p_ratio/((Mol_before+Mol_after)/2);
            % Ratio C1 part in representative velocity
            Velocity_C1=1;
            % Ratio C2 part in representative velocity
            Velocity_C2=p_ratio_cmb/Mol_before;
            % Scaling vector of the two parts of the velocity
            scaling_C2Part=weight_C2Part.*Velocity_C2;
            scaling_C1Part=weight_C1Part.*Velocity_C1;
            % Complete scaling vector of the representative velocity
            scaling_rep_vel=scaling_C2Part+scaling_C1Part;
            % "weight_C1Part" and "weight_C2Part"
            % should be an approximation to Fig. 4 b):
            % (1xN) matrix

            % Size of weighting vector for alpha individual ratios
            weight_size=size(xi_rel);
            weight_size=weight_size(1,2);
            % Transformation scaling factor
            beta=ones(1,weight_size);
            beta=beta.*(p_ratio.^m_used);
            beta=beta.*(T_ratio.^(0.75-1.62*m_used));
            beta=beta.*(scaling_rep_vel.^m_used);
            f_tilde=f_tilde.*beta;
        elseif Cycle==4 % Fourth engine stroke
            % Straight forward
        end
end
```

Bibliography

Adair, R. P., Qvale, E. B., Pearson, J. T., 1972. Instantaneous heat transfer to cylinder wall in reciprocating compressors. International Compressor Engineering Conference Purdue University 86, 521–526.

Ananthraraman, S., Baskaran, M., 2015. A Research on Factors Affecting Nozzle Tip Temperature in Diesel Engines. In: SAE Tech. Pap. 2015-01-2791. SAE international.

Annand, W. J. D., 1963. Heat Transfer in the Cylinders of Reciprocating Internal Combustion Engines. Proc. Inst. Mech. Eng. 177 (1), 973–996.

Baehr, H., Stephan, K., 2010a. Wärme und Stoffübertragung, 7th Edition. Springer-Verlag, Berlin Heidelberg New York.

Baehr, H. D., Stephan, K., 2010b. Wärme und Stoffübertragung, 7th Edition. Springer-Verlag, Berlin Heidelberg New York.

Banjac, T., Wurzenberger, J. C., Katrasnik, T., 2014. Assessment of engine thermal management through advanced system engineering modeling. Adv. Eng. Softw. 71, 19–33.

Bargende, M., 1991. Ein Gleichungssatz zur Berechnung der instationären Wandwärmeverluste im Hochdruckteil von Ottomotoren. Doctoral dissertation, Technical University of Darmstadt.

Bargende, M., Heinle, M., Berner, H.-J., 2011. Einige Ergänzungen zur Berechnung der Wandwärmeverluste in der Prozessrechnung. In: 13th Symp. Work. Process Intern. Combust. Engine. Graz.

Bauer, W. D., Wenisch, J., Heywood, J. B., 1998. Averaged and time-resolved heat transfer of steady and pulsating entry flow in intake manifold of a spark-ignition engine. Int. J. Heat Fluid Flow 19 (1), 1–9.

Bäumel, F., 2001. Werkstoffgerechte Auslegung und Festigkeitsnachweis für Verbrennungsmotorkolben aus Mesophasenkohlenstoff. Doctoral dissertation, Technical University of Munich.

Berni, F., Cicalese, G., Fontanesi, S., 2017. A modified thermal wall function for the estimation of gas-to-wall heat fluxes in CFD in-cylinder simulations of high performance spark-ignition engines. Appl. Therm. Eng. 115, 1045–1062.

Boelter, L. M. K., Cherry, V. H., Johnson, H. A., Martinelli, R., 1965. Heat Transfer Notes. McGraw-Hill Book Company, New York.

Borgnakke, C., Arpaci, V. S., Tabaczynski, R. J., 1980. A Model for the Instantaneous Heat Transfer and Turbulence in a Spark Ignition Engine. In: SAE Tech. Pap. 800287. SAE International.

Bozza, F., Teodosio, L., De Bellis, V., Fontanesi, S., Iorio, A., 2018. Refinement of a 0D Turbulence Model to Predict Tumble and Turbulent Intensity in SI Engines. Part II: Model Concept, Validation and Discussion. In: SAE Tech. Pap. 2018-01-0856.

Bridgman, P. W., 1932. Theorie der physikalischen Dimensionen. Teubner Verlag Berlin.

Broekaert, S., Cuyper, T. D., Paepe, M. D., Verhelst, S., 2017. Evaluation of empirical heat transfer models for HCCI combustion in a CFR engine. Appl. Energy 205, 1141–1150.

Broekaert, S., Demuynck, J., Cuyper, T. D., Paepe, M. D., Verhelst, S., 2016. Heat transfer in premixed spark ignition engines part I : Identification of the factors influencing heat transfer. Energy 116, 380–391.

Caton, J. A., Heywood, J. B., 1981. An experimental and analytical study of heat transfer in an engine exhaust port. Int. J. Heat Mass Transf. 24 (4), 581–595.

Chaboche, J. L., 1989. Constitutive equations for cyclic plasticty and cyclic viscoplasticity. Int. J. Plast. 5, 247–302.

Chaboche, J. L., 2008. A review of some plasticity and viscoplasticity constitutive theories. Int. J. Plast. 24 (10), 1642–1693.

Chang, J., Güralp, O., Filipi, Z., Assanis, D. N., Kuo, T.-W., Najt, P., Rask, R., 2004. New Heat Transfer Correlation for an HCCI Engine Derived from Measurements of Instantaneous Surface Heat Flux. In: SAE Tech. Pap. 2004-01-2996. SAE international.

Chen, X., Brewer, T., Sever, C., Prabhu, E., Adimi, R., Engler-pinto, C., 2017. Cylinder Head Design Process to Improve High Cycle Fatigue Performance. In: SAE Tech. Pap. 2017-01-1074.

Chiodi, M., Bargende, M., 2001. Improvement of Engine Heat-Transfer Calculation in the Three-Dimensional Simulation Using a Phenomenological Heat-Transfer Model. In: SAE Tech. Pap. 2001-01-3601. SAE international.

Choi, G. H., Choi, K. H., Lee, J. T., Song, Y. S., Ryu, Y., Cho, J. W., 1997. Analysis of Combustion Chamber Temperature and Heat Flux in a DOHC Engine. In: SAE Tech. Pap. 970895. SAE International.

Colburn, A. P., 1964. A method of correlating forced convection heat transfer data and a comparison with fluid friction. Int. J. Heat Mass Transf. 7 (12), 1359–1384.

Cuyper, T. D., Demuynck, J., Broekaert, S., Paepe, M. D., Verhelst, S., 2016. Heat transfer in premixed spark ignition engines part II : Systematic analysis of the heat transfer phenomena. Energy 116, 851–860.

Danisch, R., 2015. Motorsport expertise for top performance in everyday life. Mahle Press Release , September issue.

De Bellis, V., Bozza, F., Fontanesi, S., Severi, E., Berni, F., apr 2016. Development of a Phenomenological Turbulence Model through a Hierarchical 1D/3D Approach Applied to a VVA Turbocharged Engine. SAE Int. J. Engines 9 (1), 506–519.

De Bellis, V., Severi, E., Fontanesi, S., Bozza, F., 2014. Hierarchical 1D/3D approach for the development of a turbulent combustion model applied to a VVA turbocharged engine. Part I: turbulence model. Energy Procedia 45, 829–838.

Demuynck, J., De Paepe, M., Huisseune, H., Sierens, R., Vancoillie, J., Verhelst, S., 2011. Investigation of the influence of engine settings on the heat flux in a hydrogen- and methane-fueled spark ignition engine. Appl. Therm. Eng. 31 (6-7), 1220–1228.

Demuynck, J., De Paepe, M., Sileghem, L., Vancoillie, J., Verhelst, S., Chana, K., apr 2012. Applying Design of Experiments to Determine the Effect of Gas Properties on In-Cylinder Heat Flux in a Motored SI Engine. SAE Int. J. Engines 5 (3), 1286–1299.

Dent, J. C., Sulaiman, S. J., 1977. Convective and Radiative Heat Transfer in a High Swirl Direct Injection Diesel Engine. In: SAE Tech. Pap. 770407. SAE International.

Desmorat, R., Du Tertre, A., Gaborit, P., 2015. Multiaxial Haigh diagrams from incremental two scale damage analysis. Aerosp. Lab J. (9), 1–15.

Dileep, M., Sanjay, P. S., Mandloi, R. K., 2016. Analytical Study of Fatigue Failure of Aluminium Alloy Piston in IC Engines. Int. Res. J. Eng. Technol. 3 (4), 1665–1669.

Dowling, N. E., apr 2004. Mean Stress Effects in Stress-Life and Strain-Life Fatigue. In: SAE Tech. Pap. 2004-01-2227. SAE International.

Doyle, W., 1964. Development of Strong Aluminium Alloys. Aircr. Eng. Aerosp. Technol. 37 (7), 214–217.

Easter, J., Jarrett, C., Pespisa, C., Liu, Y. C., Alkidas, A. C., Guessous, L., Sangeorzan, B. P., sep 2014. An Area-Average Correlation for Oil-Jet Cooling of Automotive Pistons. J. Heat Transfer 136 (12), 124501–124504.

Eiamsa-Ard, S., Promvonge, P., 2010. Thermal characteristics in round tube fitted with serrated twisted tape. Appl. Therm. Eng. 30 (13), 1673–1682.

Eichelberg, G., 1939. Some investigations on old combustion-engine problems. Parts I and II. Engineering 148, 463–466, 547–550.

Eifler, W., Schlücker, E., Spicher, U., Will, G., 2009. Küttner Kolbenmaschinen, 7th Edition. Vieweg+Teubner, Wiesbaden.

Enomoto, Y., Furuhama, S., 1985. Study on Thin Film Thermocouple for Measuring Instantaneous Temperature on Surface of Combustion Chamber Wall in Internal Combustion Engine. Bull. JSME 28 (235), 108–116.

Enomoto, Y., Furuhama, S., 1986a. Heat Loss into Combustion Chamber Wall of 4-stroke Gasoline Engine : 2nd Report, Heat Loss into Cylinder Head, Intake and Exhaust Valves. Bull. JSME 29 (253), 2196–2203.

Enomoto, Y., Furuhama, S., 1986b. Study on Thin Film Thermocouple Measuring Instantaneous Temperature on Surface of Combustion Chamber Wall In Internal Combustion Engine - 2nd Report, Study on Thin Film Thermocouples Embedded in Combustion Chamber Wall. Bull. JSME 29 (256), 3434–3441.

Enomoto, Y., Furuhama, S., 1989. A Study of the Local Heat Transfer Coefficient on the Combustion Chamber Walls of a Four-Stroke Gasoline Engine. JSME Int. Journal. Ser. 2 32 (1), 107–114.

Enomoto, Y., Furuhama, S., Minakami, K., 1985. Heat Loss to Combustion Chamber Wall of 4-Stroke Gasoline Engine: 1st Report, Heat Loss to Piston and Cylinder. Bull. JSME 28 (238), 647–655.

Esfahanian, V., Javaheri, A., Ghaffarpour, M., 2006. Thermal analysis of an SI engine piston using different combustion boundary condition treatments. Appl. Therm. Eng. 26 (2-3), 277–287.

Eurocode, 1999. Eurocode 9 - Design of aluminium structures. Tech. rep., European Committee for Standardization.

Fedelich, B., Kühn, H.-j., Rehmer, B., Skrotzki, B., 2017. Experimental and analytical investigation of the TMF-HCF lifetime behavior of two cast iron alloys. Int. J. Fatigue 99, 266–278.

Federation Internationale de l'Automobile - FIA, 2018a. Formula one technical regulations.

Federation Internationale de l'Automobile - FIA, 2018b. World Endurance Championship technical regulations.

Ferraro, C. V., Marzano, M. R., Millo, F., Bochicchio, N., 1996. Comparison Between Heat Transfer and Knock Intensity on a Statistical Basis. In: SAE Tech. Pap. 962101. SAE International.

Ferziger, J. H., Peric, M., 2002. Computational Methods for Fluid Dynamics. Springer-Verlag, Berlin Heidelberg.

Fieberg, C., Korthäuer, M., 2006. Kontaktwärmeübergang - Vorhaben Nr. 828, Kontaktdruckabhängiger Wärmeübergang im motorischen Umfeld. Forschungsvereinigung Verbrennungskraftmaschinen 821.

Finol, C. A., Robinson, K., 2006. Thermal modelling of modern engines: a review of empirical correlations to estimate the in-cylinder heat transfer coefficient. Proc. Inst. Mech. Eng. Part D J. Automob. Eng. 220 (12), 1765–1781.

Fischer, C., Schweizer, C., Seifert, T., 2015. Assessment of fatigue crack closure under in-phase and out-of-phase thermomechanical fatigue loading using a temperature dependent strip yield model. Int. J. Fatigue 78, 22–30.

Fišer, J., Zítek, P., Vyhlídal, T., 2015. Magnitude Optimum Design of PID Control Loop with Delay. IFAC-PapersOnLine 48 (12), 446–451.

Fogla, N., Bybee, M., Mirzaeian, M., Millo, F., Wahiduzzaman, S., 2017. Development of a K-k-epsilon Phenomenological Model to Predict In-Cylinder Turbulence. SAE Int. J. Engines 10 (2), 562–575.

Fontanesi, S., Cicalese, G., D'Adamo, A., Pivetti, G., sep 2011. Validation of a CFD Methodology for the Analysis of Conjugate Heat Transfer in a High Performance SI Engine. In: SAE Tech. Pap. 2011-24-0132. SAE International.

Fontanesi, S., Cicalese, G., Tiberi, A., apr 2013. Combined In-cylinder/CHT Analyses for the Accurate Estimation of the Thermal Flow Field of a High Performance Engine for Sport Car Applications. In: SAE World Congr. Exhib. SAE International.

Fontanesi, S., Giacopini, M., 2013. Multiphase CFD-CHT optimization of the cooling jacket and FEM analysis of the engine head of a V6 diesel engine. Appl. Therm. Eng. 52 (2), 293–303.

Forschungskuratorium, Maschinenbau, 2005. FKM-Richtlinie, Rechnerischer Festigkeitsnachweis für Maschinenbauteile, 5th Edition. VDMA - Verlag, Frankfurt/Main.

Frost, H. J., Ashby, M. F., n.d. Deformation-Mechanism Maps - The Plasticity and Creep of Metals and Ceramics. Retrieved June 10, 2017. URL `http://engineering.dartmouth.edu/defmech/`

Gamma Technologies, 2016. GT-Power User Manual, v7.5.

Ghebru, D., 2013. Modellierung und Analyse des instationären thermischen Verhaltens von Verbrennungsmotor und Gesamtfahrzeug. Doctoral dissertation, Karlsruhe Institute of Technology (KIT).

Gilaber, P., Pinchon, P., feb 1988. Measurements and Multidimensional Modeling of Gas-Wall Heat Transfer in a S.I. Engine. In: SAE Tech. Pap. 880516. SAE International.

Gnielinski, V., 1976. New equations for heat and mass transfer in turbulent pipe and channel flow. Int. Chem. Eng 16 (2), 359–368.

Gocmez, T., Pischinger, S., 2010. A contribution to the understanding of thermomechanical fatigue sensitivities in combustion engine cylinder heads. J. Automob. Eng. 225, 461–477.

Golloch, R., 2005. Downsizing bei Verbrennungsmotoren, 1st Edition. Springer-Verlag, Berlin Heidelberg.

Grieb, M. B., Christ, H.-j., Plege, B., 2010. Thermomechanical fatigue of cast aluminium alloys for cylinder head applications - experimental characterization and life prediction. Procedia Eng. 2 (1), 1767–1776.

Grill, M., 2006. Objektorientierte Prozessrechnung von Verbrennungsmotoren. Doctoral dissertation, University of Stuttgart.

Grill, M., Billinger, T., Bargende, M., apr 2006. Quasi-Dimensional Modeling of Spark Ignition Engine Combustion with Variable Valve Train. In: SAE Tech. Pap. 2006-01-1107. SAE International.

Habib, M. A., Attya, A. M., Said, S. A. M., Eid, A. I., Aly, A. Z., 2004. Heat transfer characteristics and Nusselt number correlation of turbulent pulsating pipe air flows. Heat Mass Transf. 40, 307–318.

Hadzic, I., 1999. Second-moment closure modeling of transitional and unsteady turbulent flows. Doctoral dissertation, Technical University of Delft.

Haghighat, A. K., Roumi, S., Madani, N., Bahmanpour, D., Olsen, M. G., 2018. An intelligent cooling system and control model for improved engine thermal management. Appl. Therm. Eng. 128, 253–263.

Haibach, E., 2006. Betriebsfestigkeit. Springer-Verlag, Berlin Heidelberg.

Hänel, B., Wirthgen, G., 1994. Festigkeitsnachweis. Vorhaben Nr. 154. Teil 1. Forschungskuratorium Maschinenbau, VDMA - Verlag, Frankfurt/Main.

Hasan, B. O., 2007. Turbulent Prandtl Number and its Use in Prediction of Heat Transfer Coefficient for Liquids. Coll. Eng. J. 10 (1), 53–64.

Hettinger, A., Kulzer, A., 2009. A New Method to Detect Knocking Zones. SAE Int. J. Engines 2 (1), 645–665.

Heywood, J. B., 1988. Internal Combustion Engine Fundamentals. Mcraw-Hill, New York (amongst others).

Hohenberg, G. F., 1979. Advance approaches for heat transfer calculation. In: SAE Tech. Pap. 790825.

Hölz, P., 2015. Thermische Untersuchung einer Radialturbine zur Abgasenergierückgewinnung. Master thesis, Institute of Fusion and Reactor Technology (KIT).

Hölz, P., Böhlke, T., Krämer, T., 2018a. CFD-CHT calculation method using Buckingham Pi-Theorem for complex fuid-solid heat transfer problems with scattering boundary conditions. Automot. Engine Technol. , 1–16.

Hölz, P., Böhlke, T., Krämer, T., 2018b. Fast algorithms for generating thermal boundary conditions in combustion chambers. Appl. Therm. Eng. 141, 101–113.

Hölz, P., Böhlke, T., Krämer, T., 2019a. Determining water mass flow control strategies for a turbocharged SI engine using a two-stage calculation method. Appl. Therm. Eng. 146 (5), 386–395.

Hölz, P., Böhlke, T., Krämer, T., 2019b. Transient temperature calculation method for complex fluid-solid heat transfer problems with scattering boundary conditions. Appl. Therm. Eng. 149, 1463–1475.

Hörmann, T., Lechner, B., Puntigam, W., Moshammer, T., Almbauer, R., 2005. Numerical and Experimental Investigation of Flow and Temperature Fields around Automotive Cooling Systems. In: SAE Tech. Pap. 2005-01-2006. SAE International.

Huegel, P., Kubach, H., Koch, T., Velji, A., 2015. Investigations on the Heat Transfer in a Single Cylinder Research SI Engine with Gasoline Direct Injection. SAE Int. J. Engines 8 (2), 557–569.

Irimescu, A., 2013. Convective heat transfer equation for turbulent flow in tubes applied to internal combustion engines operated under motored conditions. Appl. Therm. Eng. 50 (1), 536–545.

JANAF, 1971. Thermochemical Tables, 2nd Edition. NSRDS-NB537, U.S. National Bureau of Standards.

Jiang, Y., Feng, M., 2004. Modeling of Fatigue Crack Propagation. J. Eng. Mater. Technol. 126 (1), 77.

Karnik, A., Pachner, D., Fuxman, A. M., Germann, D., Jankovic, M., House, C., apr 2015. Model Predictive Control for Engine Powertrain Thermal Management Applications. In: SAE Tech. Pap. 2015-01-0336. SAE International.

Kaufman, J., 2008. Properties of Aluminium Alloys. ASM International, Ohio.

Kenjeres, S., Gunarjo, S., Hanjalic, K., 2005. Contribution to elliptic relaxation modelling of turbulent natural and mixed convection. Int. J. Heat Fluid Flow 26 (4), 569–586.

Kenningley, S., Morgenstern, R., 2012. Thermal and Mechanical Loading in the Combustion Bowl Region of Light Vehicle Diesel AlSiCuNiMg Pistons; Reviewed with Emphasis on Advanced Finite Element Analysis and Instrumented Engine Testing Techniques. In: SAE Tech. Pap. 2012-01-1330. SAE International.

Kessler, C., 1955. Über die Vorausberechnung optimal abgestimmter Regelkreise Teil III. Die optimale Einstellung des Reglers nach dem Betragsoptimum. Regelungstechnik 3 (2), 40–49.

Kessler, C., 1958. Das symmetrische Optimum. Regelungstchnik 6, 432–436.

Khalil, O., 2014. Isothermes Kurzzeitermüdungsverhalten der hochwarmfesten Aluminium-Knetlegierung 2618A. Doctoral dissertation, Karlsruhe Institute of Technology (KIT).

Klein, M., Eriksson, L., 2004. A Specific Heat Ratio Model for Single-Zone Heat Release Models. In: SAE Tech. Pap. 2004-01-1464. SAE International.

Kleinschmidt, W., 1993. Zur Theorie und Berechnung der instationären Wärmeübertragung in Verbrennungsmotoren. In: 4. Tagung "Der Arbeitsprozess des Verbrennungsmotors". Graz.

Koch, T., 2002. Numerischer Beitrag zur Charakterisierung und Vorausberechnung der Gemischbildung und Verbrennung in einem direkteingespritzten, strahlgeführten Ottomotor. Doctoral dissertation, ETH Zürich.

Königsson, F., Stalhammar, P., Angstrom, H.-E., 2012. Controlling the Injector Tip Temperature in a Diesel Dual Fuel Engine. In: SAE Tech. Pap. 2012-01-0826. SAE international.

Kothare, M. V., Balakrishnan, V., Morari, M., 1996. Robust constrained model predictive control using linear matrix inequalities. Automatica 32 (10), 1361–1379.

Koutiri, I., Bellett, D., Morel, F., Pessard, E., 2013. A probabilistic model for the high cycle fatigue behaviour of cast aluminium alloys subject to complex loads. Int. J. Fatigue 47, 137–147.

Kozuch, P., 2004. Ein phänomenologisches Modell zur kombinierten Stickoxid- und Rußberechnung bei direkteinspritzenden Dieselmotoren. Doctoral dissertation, University of Stuttgart.

Kraus, E., 2007. Simulation der vorgemischten Verbrennung in einem realen Motor mit dem Level-Set-Ansatz. Doctoral dissertation, Eberhard-Karls-University Tübingen.

Kunkelmann, C., 2011. Numerical Modeling and Investigation of Boiling Phenomena. Doctoral dissertation, Technical University Darmstadt.

Launder, B. E., 1972. The prediciton of laminarization with a two-equation model of turbulence. Int. J. Heat Mass Transf. 15, 301–314.

Lee, T. K., Filipi, Z. S., feb 2011. Improving the predictiveness of the quasi-d combustion model for spark ignition engines with flexible intake systems. Int. J. Automot. Technol. 12 (1), 1–9.

LeFeuvre, T., Myers, P. S., Uyehara, O. A., 1969. Experimental Instantaneous Heat Fluxes in a Diesel Engine and Their Correlation. In: SAE Tech. Pap. 690464. SAE International.

Lejsek, D., Kulzer, A., Hammer, J., nov 2010. A novel transient wall heat transfer approach for the start-up of SI engines with gasoline direct injection. Heat Mass Transf. 46 (10), 1053–1067.

Lienhard, J. H. I., Lienhard, J. H. V., 2003. A heat transfer textbook, 3rd Edition. Phlogiston Press, Cambridge Massachusetts.

Liu, C., Jiang, D., Obokata, T., 2000. A Simplified Turbulence Model for In-cylinder Gas Flow in Quasi-dimensional Turbulence Combustion Model for Spark-ignition Engines. In: SAE Tech. Pap. 2000-01-2803. SAE International.

Lu, P., Gao, Q., Wang, Y., 2016. The simulation methods based on 1D-3D collaborative computing for the vehicle integrated thermal management. Appl. Therm. Eng. 104, 42–53.

Lumm, J., 1966. Mechanical Properties of 2618 Aluminium Alloy. Tech. rep., Air Force Materials Laboratory, Ohio.

Luo, X., Yu, X., Jansons, M., 2015. Simultaneous In-Cylinder Surface Temperature Measurements with Thermocouple , Laser-induced Phosphorescence , and Dual Wavelength Infrared Diagnostic Techniques in an Optical Engine. In: SAE Tech. Pap. 2005-01-1658. SAE international.

Malbec, L. M., Picket, L. M., Payri, R., Manin, J., Bardi, M., Bazin, T., Kastengren, A., Meijer, M., Somers, L. M. T., 2011. Nozzle tip and injectors body temperature measurements. In: ECN Work. 13-14 May.

Mancaruso, E., Sequino, L., Vaglieco, B., Maroteaux, F., 2018. Experimental measurements of piston temperature and evaluation of heat flux in engine at transient conditions. In: THIESEL Conf. Thermo- Fluid Dyn. Process. Direct Inject. Engines.

Mathur, S., Tondon, P., Saxena, S., 1967. Thermal conductivity of binary, ternary and quaternary mixtures of rare gases. Mol. Phys. 12, 569–579.

McBride, B., Gordon, S., Reno, M., 1993. Coefficients for Calculating Thermodynamic and Transport Properties of Individual Species. Tech. rep., NASA Report TM-4513.

Menter, F. R., 1994. Two-Equation Eddy-Viscosity Turbulence Models for Engineering Applications. AIAA J. 32 (8), 1598–1605.

Merker, G., Baumgarten, C., 2000. Fluid- und Wärmetransport Strömungslehre. Vieweg+Teubner Verlag.

Merker, G. P., Teichmann, R., 2014. Grundlagen Verbrennungsmotoren. Springer Fachmedien, Wiesbaden.

Metzger, M., Leidenfrost, M., Werner, E., Riedel, H., Seifert, T., 2014. Lifetime Prediction of EN-GJV 450 Cast Iron Cylinder Heads under Combined Thermo-Mechanical and High Cycle Fatigue Loading. SAE Int. J. Engines 7 (2).

Metzger, M., Seifert, T., 2012. A mechanism-based model for LCF/HCF and TMF/HCF life prediction: Multiaxial formulation, finite-element implementation and application to cast iron. Tech. Mech. 32 (2), 435–445.

Michels, K., Hoffmeyer, H., Theobald, J., 2012. Ermittlung der thermischen Randbedingungen an einem aufgeladenen direkteinspritzenden Ottomotor mit Oberflächenthermoelementen. In: 10th Int. Symp. Combust. Diagnostic.

Michl, J., Neumann, J., Rottengruber, H., Wensing, M., 2016. Derivation and validation of a heat transfer model in a hydrogen combustion engine. Appl. Therm. Eng. 98, 502–512.

Mizuno, H., Ashida, K., Teraji, A., Ushijima, K., Takemura, S., 2009. Transient Analysis of the Piston Temperature with Consideration of In-cylinder Phenomena Using Engine Measurement and Heat Transfer Simulation Coupled with Three-dimensional Combustion Simulation. SAE Int. J. Engines 2 (1), 83–90.

Mohamed, E. S., 2016. Experimental study on the effect of active engine thermal management on a bi-fuel engine performance, combustion and exhaust emissions. Appl. Therm. Eng. 106, 1352–1365.

Mohammadi, A., Yaghoubi, M., 2010. Estimation of instantaneous local heat transfer coefficient in spark-ignition engines. Int. J. Therm. Sci. 49 (7), 1309–1317.

Mohammadi, A., Yaghoubi, M., 2011. Two Phase Flow Simulation for Subcooled Nucleat Boiling Heat Transfer Calculation in Water Jacket of Diesel Engine. J. Engine Res. 22, 50–61.

Mohammadi, A., Yaghoubi, M., Rashidi, M., 2008. Analysis of local convective heat transfer in a spark ignition engine. Int. Commun. Heat Mass Transf. 35 (2), 215–224.

Morris, W. L., Buck, O., Marcus, H. L., 1976. Fatigue crack initiation and early propagation in Al 2219-T851. Met. Trans. A 7 (7), 1161–1165.

Moschandreou, T., Zamir, M., 1997. Heat transfer in a tube with pulsating flow and constant heat flux. Int. J. Heat Mass Transf. 40 (10), 2461–2466.

Myers, M. R., 1990. Damage Accumulation During High Temperature Testing of a Squeeze Formed Aluminum Piston Alloy. J. Mater. Manuf. 99, 99–106.

Nallasamy, M., 1987. Turbulence models and their applications to the prediction of internal flows: A review. Comput. Fluids 15 (2), 151–194.

Nicouleau, E., Feyel, F., Quilici, S., Cailletaud, G., 2002. Structural calculation and lifetime-prediction in thermomechanical fatigue of engine components. In: Temp. Interact. Vol. 29 of European Structural Integrity Society. pp. 331–340.

Oehrby, F., 2014. Numerical modeling of subcooled nucleate flow boiling in engine cooling systems. Doctoral dissertation, Chalmers University of Technology.

Oertel jr., H., Böhle, M., Reviol, T., 2011. Strömungsmechanik. Vieweg+Teubner Verlag, Wiesbaden.

Pallarés-Santasmartas, L., Albizuri, J., Avilés, A., Avilés, R., 2018. Mean Stress Effect on the Axial Fatigue Strength of DIN 34CrNiMo6 Quenched and Tempered Steel. Metals (Basel). 8 (4), 1–19.

Pang, H. H., Brace, C. J., 2004. Review of engine cooling technologies for modern engines. Proc. Inst. Mech. Eng. Part D J. Automob. Eng. 218 (11), 1209–1215.

Papadopoulos, I. V., 1996. Exploring the High-Cycle Fatigue Behaviour of Metals from the Mesoscopic Scale. J. Mech. Behav. Mater. 6 (2), 93–118.

Papadopoulos, I. V., Davoli, P., Gorla, C., Filippini, M., Bernasconi, A., 1997. A comparative study of multiaxial high-cycle fatigue criteria for metals. Int. J. Fatigue 19 (3), 219–235.

Patel, J. T., Attal, M. H., 2016. An Experimental Investigation of Heat Transfer Characteristics of Pulsating Flow in Pipe. Int. J. Curr. Eng. Technol. 6 (5), 1515–1521.

Payri, F., Margot, X., Gil, A., Martin, J., 2005. Computational Study of Heat Transfer to the Walls of a DI Diesel Engine. In: SAE Tech. Pap. 2005-01-0210. SAE International.

Peralta, P., Laird, C., 2014. Fatigue of Metals. In: Phys. Metall., 5th Edition. Elsevier, pp. 1765–1880.

Petukhov, B. S., 1970. Heat Transfer and Friction in Turbulent Pipe Flow with Variable Physical Properties. Adv. Heat Transf. 6, 504–564.

Pizzonia, F., Castiglione, T., Bova, S., 2016. A Robust Model Predictive Control for efficient thermal management of internal combustion engines. Appl. Energy 169, 555–566.

Pope, S. B., 2010. Turbulent Flows, 7th Edition. Cambridge University Press.

Punekar, H., Das, S., apr 2013. Numerical Simulation of Subcooled Nucleate Boiling in Cooling Jacket of IC Engine. In: SAE Tech. Pap. 2013-01-1651. SAE International.

Radaj, D., Vormwald, M., 2007. Ermüdungsfestigkeit. Springer-Verlag, Berlin Heidelberg.

Rakopoulos, C., Mavropoulos, G., 1996. Study of the steady and transient temperature field and heat flow in the combustion chamber components of a medium speed diesel engine using finite element analyses. Int. J. Energy Res. 20, 437–464.

Reichardt, H., 1951. Vollständige Darstellung der turbulenten Geschwindigkeitsverteilung in glatten Leitungen. Z. Angew. Math. Mech 31 (7), 208–219.

Robinson, J. S., Cudd, R. L., Evans, J. T., 2003. Creep resistant aluminium alloys and their applications. Mater. Sci. Technol. 19 (2), 143–155.

Rodi, W., 1991. Experience with two-layer models combining the k-epsilon model with a one-equation model near the wall. AIAA 91-0216.

Rodi, W., 2017. Turbulence Modeling and Simulation in Hydraulics : A Historical Review. J. Hydraul. Eng. 143 (5), 1–20.

Sangid, M. D., 2013. The physics of fatigue crack initiation. Int. J. Fatigue 57, 58–72.

Sanli, A., Ozsezen, A. N., Kilicaslan, I., Canekci, M., 2008. The influence of engine speed and load on the heat transfer between gases and in-cylinder walls at fired and motored conditions of an IDI diesel engine. Appl. Therm. Eng. 28, 1395–1404.

Sanli, A., Sayin, C., Gumus, M., Kilicaslan, I., Canakci, M., 2009. Numerical Evaluation by Models of Load and Spark Timing Effects on the In-Cylinder Heat Transfer of a SI Engine. Numer. Heat Transf. Part A Appl. 56 (5), 444–458.

Santecchia, E., Hamouda, A. M. S., Musharavati, F., Zalnezhad, E., Cabibbo, M., Mehtedi, M. E., Spigarelli, S., 2016. A Review on Fatigue Life Prediction Methods for Metals. Adv. Mater. Sci. Eng. 2016.

Santus, C., Taylor, D., 2009. Physically short crack propagation in metals during high cycle fatigue. Int. J. Fatigue 31 (8-9), 1356–1365.

Schubert, C., Wimmer, A., Chmela, F., 2005. Advanced Heat Transfer Model for CI Engines. In: SAE Tech. Pap. 2005-01-0695. SAE international.

Schweizer, C., Seifert, T., Nieweg, B., Von Hartrott, P., Riedel, H., 2011. Mechanisms and modelling of fatigue crack growth under combined low and high cycle fatigue loading. Int. J. Fatigue 33 (2), 194–202.

Schwiderik, V., 2013. Entwicklung einer Methode zur Bestimmung des Temperaturfeldes eines Verbrennungsmotors in seiner Konzeptphase. Master thesis, Fakultät für Technik 1 - Mechanik und Elektronik (Hochschule Heilbronn).

Seifert, T., Maier, G., Uihlein, A., Lang, K. H., Riedel, H., 2010. Mechanism-based thermomechanical fatigue life prediction of cast iron. Part II: Comparison of model predictions with experiments. Int. J. Fatigue 32 (8), 1368–1377.

Seifert, T., Riedel, H., 2009. Fatigue Life Prediction of High Temperature Components in Combustion Engines and Exhaust Systems. In: Proc. 4Th EASC.

Seifert, T., Riedel, H., 2010. Mechanism-based thermomechanical fatigue life prediction of cast iron. Part I: Models. Int. J. Fatigue 32 (8), 1358–1367.

Sellars, J., Tribus, M., Klein, J., 1954. Heat Transfer To Laminar Flow In A Round Tube Or Flat Conduit: The Graetz Problem Extended. WADC Tech. Rep. 54-255 , 1–23.

Setlur, P., Wagner, J., Dawson, D., Chen, J., 2003. Nonlinear controller for automotive thermal management systems. In: Am. Control Conf. IEEE.

Sharief, A., Chandrashekar, T. K., Antony.A.J, Samaga, B. S., 2008. Study on Heat Transfer Correlation in IC Engines. In: SAE Tech. Pap. 2008-01-1816. SAE international.

Shariyat, M., Fathi Sola, J., Jazayeri, S., 2016. Experimentally Validated Combustion and Piston Fatigue Life Evaluation Procedures for the Bi-Fuel Engines , Using an Integral-Type Fatigue Criterion. Lat. Am. J. Solids Struct. 13, 1030–1053.

Shojaefard, M. H., Noorpoor, A. R., Bozchaloe, D. A., Ghaffarpour, M., 2005. Transient Thermal Analysis of Engine Exhaust Valve. Numer. Heat Transf. Part A Appl. 48 (7), 627–644.

Shudo, T., Nakajima, Y., Futakuchi, T., 2000. Thermal efficiency analysis in a hydrogen premixed combustion engine. JSAE Rev. 21 (2), 177–182.

Siemens, 2016. STAR-CCM+ User Manual, v11.06.

Sihling, K., Woschni, G., 1979. Experimental Investigation of the Instantaneous Heat Transfer in the Cylinder of a High Speed Diesel Engine. In: SAE Tech. Pap. 790833. SAE International.

Silva, F. S., 2006. Fatigue on engine pistons - A compendium of case studies. Eng. Fail. Anal. 13, 480–492.

Soyhan, H. S., Yasar, H., Walmsley, H., Head, B., Kalghatgi, G. T., Sorusbay, C., 2009. Evaluation of heat transfer correlations for HCCI engine modeling. Appl. Therm. Eng. 29 (2-3), 541–549.

Su, X., Zubeck, M., Lasecki, J., Jr, C. C. E.-p., Tang, C., Allison, J., 2002. Thermal Fatigue Analysis of Cast Aluminum Cylinder Heads. In: SAE Tech. Pap. 2002-01-0657. SAE International.

Sugihara, T., Suzuki, Y., Shimano, K., Enomoto, Y., Emi, M., 2007. Direct Heat Loss to Combustion Chamber Walls in a D.I. Diesel Engine. In: SAE Tech. Pap. 2007-24-0006. SAE international.

Taler, D., Taler, J., 2017. Simple heat transfer correlations for turbulent tube flow. In: E3S Web Conf. Vol. 13.

Tanaka, K., Mura, T., mar 1981. A Dislocation Model for Fatigue Crack Initiation. J. Appl. Mech. 48 (1), 97–103.

Tanaka, K., Mura, T., 1982a. A micromechanical theory of fatique crack initiation from notches. Mech. Mater. 1 (1), 63–73.

Tanaka, K., Mura, T., 1982b. A theory of fatigue crack initiation at inclusions. Met. Trans. A 13 (1), 117–123.

Tanaka, K., Nakai, Y., Yamashita, M., 1981. Fatigue growth threshold of small cracks. Int. J. Fract. 17 (5), 519–533.

Taylor, C. F., Toong, T. Y., 1957. Heat Transfer in Internal-combustion Engines. ASME Pap. 57-HT-17 .

Tillock, B. R., Martin, J. K., 1996. Measurement and modeling of thermal flows in an air-cooled engine. In: SAE Tech. Pap. 961731. SAE International.

Toninel, S., Forkel, H., Frank, T., Durst, B., Hasse, C., Linse, D., apr 2009. Implementation and Validation of the G-equation Model Coupled with Flamelet Libraries for Simulating Premixed Combustion in I.C. Engines. SAE Int. J. Engines 2 (1), 674–690.

Uchida, K., Suzuki, T., Oguri, Y., Yoshida, M., 2005. Precise Measurement of Heat Transfer to the Inlet Air using Intake Port Model. In: SAE Tech. Pap. 2005-01-0999. SAE International.

Varghese, M. B., Agarwal, A. K., 2004. Numerical Investigations Of Piston Cooling Using Oil Jet. In: SAE Tech. Pap. 2004-28-0061.

Voda, A. A., Landau, I. D., 1995. A method for the auto-calibration of PID controllers. Automatica 31 (1), 41–53.

Wagner, J. R., Srinivasan, V., Dawson, D. M., Marotta, E. E., 2003. Smart Thermostat and Coolant Pump Control for Engine Thermal Management Systems. In: SAE Tech. Pap. 2003-01-0272. SAE International.

Walliser, F.-S., 2015. Motorsports as a driver of new technologies. Porsche Press Release , Nr. 110/15.

Wang, C., Shang, D. G., Wang, X. W., 2014. A New Multiaxial High-Cycle Fatigue Criterion Based on the Critical Plane for Ductile and Brittle Materials. J. Mater. Eng. Perform. 24 (2), 816–824.

Wang, C. S., Berry, G. F., 1985. Heat Transfer in Internal Combustion Engines. In: ASME Winter Annu. Meet.

Wang, G., Gao, Q., Zhang, T., Wang, Y., 2016. A simulation approach of under-hood thermal management. Adv. Eng. Softw. 100, 43–52.

Wang, X., Price, P., Stone, C. R., Richardson, D., 2007. Heat release and heat flux in a spray-guided direct-injection gasoline engine. Proc Inst Mech Eng Part D J Automob Eng 221, 1441–1452.

Warnatz, J., Maas, U., Dibble, R. W., 2006. Combustion, 4th Edition. Springer-Verlag, Berlin Heidelberg.

Warrier, G. R., Dhir, V. K., feb 2006. Heat Transfer and Wall Heat Flux Partitioning During Subcooled Flow Nucleate Boiling - A Review. J. Heat Transfer 128 (12), 1243–1256.

Wimmer, A., Pivec, R., Sams, T., 2000. Heat Transfer to the Combustion Chamber and Port Walls of IC Engines - Measurement and Prediction. SAE Tech. Pap. 2000-01-0568 .

Wolfstein, M., 1969. The velocity and temperature distribution in one-dimensional flow with turbulence augmentation and pressure gradient. Int. J. Heat Mass Transf. 12, 301–318.

Wolgast, E. J., 2014. Ein Beitrag zur Prozessrechnung an Verbrennungsmotoren unter Berücksichtigung der realen Gaseigenschaften. Doctoral dissertation, Helmut-Schmidt-University Hamburg.

Woschni, G., 1967. A Universally Applicable Equation for the Instantaneous Heat Transfer Coefficient in the Internal Combustion Engine. In: SAE Tech. Pap. 670931.

Woschni, G., 1979. Prediction of Thermal Loading of Supercharged Diesel Engines. In: SAE Tech. Pap. 790821. SAE International.

Wunderlich, W., Hayashi, M., 2012. Thermal Cyclic Fatigue Analysis of Three Aluminum Piston Alloys. Int. J. Mater. Mech. Eng. 1, 57–60.

Yang, J., 1996. Coolant Pump Throttling - A Simple Method to Improve the Control Over SI Engine Cooling System. In: SAE Tech. Pap. 961813. SAE International.

Yang, L. C., Hamada, A., Ohtsubo, K., 2000. Engine Valve Temperature Simulation System. In: SAE Tech. Pap. 2000-01-0564. SAE International.

Yildirim, U., Mutlu, A., Söylemez, M. T., 2015. An Optimal Robust Tuning Method for First Order plus Dead Time Systems with Parameter Uncertainty. IFAC-PapersOnLine 48 (14), 126–131.

Yokomori, T., 2017. Super-Lean Burn Technology for High Thermal Efficiency SI Engines. In: FVV 2017 Autumn Conf. Leipzig.

Zhang, G., Li, B., Zhang, J., Feng, Z., Wei, Z., Cai, W., 2012. Unique cyclic deformation behavior of a heavily alloyed Al - Si piston alloy at different temperatures. Prog. Nat. Sci. - Mater. Int. 22 (5), 445–451.

Zhao, L., Moiz, A. A., Som, S., Fogla, N., Bybee, M., Wahiduzzaman, S., Mirzaeian, M., Millo, F., Kodavasal, J., 2018. Examining the role of flame topologies and in-cylinder flow fields on cyclic variability in spark-ignited engines using large-eddy simulation. Int. J. Engine Res. 19 (8), 886–904.

Zhao, T., Jiang, Y., 2008. Fatigue of 7075-T651 aluminum alloy. Int. J. Fatigue 30 (5), 834–849.

Ziegler, J. G., Nichols, N. B., 1995. Optimum settings for automatic controllers. InTech 42 (6), 94–100.

Schriftenreihe Kontinuumsmechanik im Maschinenbau
Karlsruher Institut für Technologie (KIT)
(ISSN 2192-693X)

Herausgeber: Prof. Dr.-Ing. Thomas Böhlke

Band 1 Felix Fritzen
Microstructural modeling and computational homogenization of the physically linear and nonlinear constitutive behavior of micro-heterogeneous materials. 2011
ISBN 978-3-86644-699-1

Band 2 Rumena Tsotsova
Texturbasierte Modellierung anisotroper Fließpotentiale. 2012
ISBN 978-3-86644-764-6

Band 3 Johannes Wippler
Micromechanical finite element simulations of crack propagation in silicon nitride. 2012
ISBN 978-3-86644-818-6

Band 4 Katja Jöchen
Homogenization of the linear and non-linear mechanical behavior of polycrystals. 2013
ISBN 978-3-86644-971-8

Band 5 Stephan Wulfinghoff
Numerically Efficient Gradient Crystal Plasticity with a Grain Boundary Yield Criterion and Dislocation-based Work-Hardening. 2014
ISBN 978-3-7315-0245-6

Band 6 Viktor Müller
Micromechanical modeling of short-fiber reinforced composites. 2016
ISBN 978-3-7315-0454-2

Die Bände sind unter www.ksp.kit.edu als PDF frei verfügbar oder als Druckausgabe bestellbar.

Band 7 Florian Rieger
Work-hardening of dual-phase steel. 2016
ISBN 978-3-7315-0513-6

Band 8 Vedran Glavas
Micromechanical Modeling and Simulation of Forming Processes. 2017
ISBN 978-3-7315-0602-7

Band 9 Eric Bayerschen
Single-crystal gradient plasticity with an accumulated plastic slip: Theory and applications. 2017
ISBN 978-3-7315-0606-5

Band 10 Bartholomäus Brylka
Charakterisierung und Modellierung der Steifigkeit von langfaserverstärktem Polypropylen. 2017
ISBN 978-3-7315-0680-5

Band 11 Rudolf Neumann
Two-Scale Thermomechanical Simulation of Hot Stamping. 2017
ISBN 978-3-7315-0714-7

Band 12 Mauricio Lobos Fernández
Homogenization and materials design of mechanical properties of textured materials based on zeroth-, first- and second-order bounds of linear behavior. 2018
ISBN 978-3-7315-0770-3

Band 13 Malte Schemmann
Biaxial Characterization and Mean-field Based Damage Modeling of Sheet Molding Compound Composites. 2018
ISBN 978-3-7315-0818-2

Band 14 Jürgen Albiez
Finite element simulation of dislocation based plasticity and diffusion in multiphase materials at high temperature. 2019
ISBN 978-3-7315-0918-9

Band 15 Maria Loredana Kehrer
Thermomechanical Mean-Field Modeling and Experimental Characterization of Long Fiber-Reinforced Sheet Molding Compound Composites. 2019
ISBN 978-3-7315-0924-0

Band 16 Peter Hölz
A dynamic and statistical analysis of the temperature- and fatigue behavior of a race power unit – The effect of different thermodynamic states. 2020
ISBN 978-3-7315-0988-2